Artificial Intelligence of Things in Smart Environments

Artificial Intelligence of Things in Smart Environments

Applications in Transportation and Logistics

Edited by
Mariyam Ouaissa, Zakaria Boulouard,
Mariya Ouaissa, Yassine Maleh

DE GRUYTER

Editors
Dr. Mariyam Ouaissa
Moulay Ismail University Meknes
BP:298 Marjane 2
Meknes 50050
Morocco
mariyam.ouaissa@edu.umi.ac.ma

Dr. Mariya Ouaissa
Moulay Ismail University Meknes
BP:298 Marjane 2
Meknes 50050
Morocco
mariya.ouaissa@edu.umi.ac.ma

Dr. Zakaria Boulouard
Hassan II University
FST Mohammedia
20000 Casablanca
Morocco
zakaria.boulouard@fstm.ac.ma

Dr. Yassine Maleh
École Nationale des Sciences Appliquées
BP 77 Boulevard Beni Amir
25000 Khouribga
Morocco
yassine.maleh@ieee.org

ISBN 978-3-11-075533-6
e-ISBN (PDF) 978-3-11-076047-7
e-ISBN (EPUB) 978-3-11-076049-1
ISSN 2626-5443

Library of Congress Control Number: 2021950669

Bibliographic information published by the Deutsche Nationalbibliothek
The Deutsche Nationalbibliothek lists this publication in the Deutsche Nationalbibliografie; detailed bibliographic data are available on the Internet at http://dnb.dnb.de.

Cover image: wacomka/iStock/Getty Images Plus
Typesetting: Integra Software Services Pvt. Ltd.
Printing and binding: CPI books GmbH, Leck

www.degruyter.com

Preface

Over the past few years, the Internet of things (IoT) has introduced the possibility to design a whole new concept of our world, "smart environments." The ability to connect every object of our surroundings to sensors and computers has offered a whole new level of welfare in our everyday lives by reducing the level of human interventions through automated machine-to-machine communications. Artificial intelligence/machine learning (AI/ML) approaches have emerged as powerful tools that can optimize the performances of every part of an IoT-based smart environment.

Our objective in this book is to propose a collection of contributions including new techniques, methods, algorithms, practical solutions, and models based on the incorporation of computational intelligence–based smart IoT into application on transportation and logistics. In addition, our book illustrates how AI can provide solid ground and address the challenges for both design and analysis of different emergent systems.

This volume focuses on the use of AI/ML-based techniques to solve issues related to IoT-based intelligent transportation systems (ITS), as well as their applications. It is a collection of original contributions regarding state-of-the-art AI/ML-based solutions for signal detection, channel modeling, routing protocol design, transport-layer optimization, user/application behavior prediction, software-defined networking, communication network optimization, and security and anomaly detection. Nine chapters in this book is an attempt to present solutions to these issues.

Chapter 1 offers an overview of the vehicle environment toward Internet of vehicles (IoV). First, they begin by presenting the main characteristics, the main types of applications, and the main components defining vehicular networks. Then, they present the types of vehicle-to-everything communication and the different access technologies used in this system. Finally, they justify the evolution of vehicular networks from an ad hoc approach toward a centralized approach, namely, the IoV and they introduce different technological solutions that could be integrated into the vehicular communication architecture to partially meet the limits of the ITS architecture.

Chapter 2 introduces the types of ad hoc networks and provides a comparative analysis of wireless communication technologies and routing protocols for vehicular ad hoc networks (VANETs). It also discusses different techniques like machine learning (ML), deep learning, metaheuristic search algorithms and optimization techniques, and their ability to enhance the capabilities of VANETs. In addition, for secure communication in VANETs, intrusion detection systems (IDS) are studied to identify different security attacks.

In order to ensure the accuracy of intrusion detection in IoV, the main purpose of Chapter 3 is to apply ML algorithms for training the IDS with a sufficient dataset of security menaces. The IDS based on ML will be able to make crucial decisions in case of intrusion detection whereas continuing to learn about their highly dynamic

https://doi.org/10.1515/9783110760477-202

environment. This chapter describes different security issues in IoV and presents diverse ML algorithms employed for constructing IDS toward protecting IoV from diverse cyber-attacks.

When it comes to self-driving autonomous cars, Chapter 4 has introduced a deep learning–based approach to optimize automated steering that recognizes the instant traffic patterns captured by the front camera placed at the front of the car dashboard.

Chapter 5 presents novel and functional deep fully convolutional neural (FCN) network architectures based on transfer learning for semantic pixel-wise segmentation. The first one is a fine-tuned FCN-based VGG 16, where the second uses this first model as an encoder and concatenates its output with a fine-tune-based Xception model, which works as a decoder. The second model breaks an input image into different coherent semantically meaningful regions. These regions are then classified into predetermined classes. Exhaustive experiments are executed on three real-world datasets that confirm the effectiveness of our models trained on large or small datasets.

Chapter 6 presents an ML-based framework for automatically detecting and diagnosing anomalous behavior in vehicles. The proposed model leverages historical data collected from various vehicle sensors to extract signatures of previously observed anomalies. By selecting the required features, the framework identifies anomalies and extracts their profiles.

Chapter 7 discusses how AI and IoT continue to revolutionize agriculture with the application of smart agriculture. AI and IoT's involvement in agriculture makes it easy to make predictions like temperature, rainfall, humidity, and soil statuses. Agriculture's AI and IoT continue to advance farming practices in the end bringing profitable ventures.

Chapter 8 covers a general, parametric, and performance analysis of four optical elements to be used in microconcentrator photovoltaic systems for electric car charging station using the same primary optic: the Fresnel lens. It concerns two nonimaging optics(pyramid and hyperbola) and two imaging optics (circular and square dome).

Chapter 9 presents a system for sentiment classification of reviews posted on social media in a way that can provide an advantage for supply chain managers. More clearly, the proposed system is a hybrid method of deep learning architectures, specifically recurrent neural networks and long short-term memory, using a word embedding Word2vec to embed the features in a review to predict the sentiment.

Contents

De Gruyter Series on the Internet of Things

Edited by
Nilanjan Dey, Parikshit Mahalle

Already published in the series

The Internet of Everything

N. Dey, G. Shinde, P. Mahalle, H. Olesen (Eds.)
ISBN 978-3-11-062548-6, e-ISBN (PDF) 978-3-11-062851-7,
e-ISBN (EPUB) 978-3-11-062578-3

Internet of Things

Kolla Bhanu Prakash (Ed.)
ISBN 978-3-11-067763-8, e-ISBN (PDF) 978-3-11-067773-7,
e-ISBN (EPUB) 978-3-11-067777-5

Artificial Intelligence of Things in Smart Environments

Artificial Intelligence of Things in Smart Environments

Applications in Transportation and Logistics

Edited by
Mariyam Ouaissa, Zakaria Boulouard,
Mariya Ouaissa, Yassine Maleh

DE GRUYTER

Editors
Dr. Mariyam Ouaissa
Moulay Ismail University Meknes
BP:298 Marjane 2
Meknes 50050
Morocco
mariyam.ouaissa@edu.umi.ac.ma

Dr. Mariya Ouaissa
Moulay Ismail University Meknes
BP:298 Marjane 2
Meknes 50050
Morocco
mariya.ouaissa@edu.umi.ac.ma

Dr. Zakaria Boulouard
Hassan II University
FST Mohammedia
20000 Casablanca
Morocco
zakaria.boulouard@fstm.ac.ma

Dr. Yassine Maleh
École Nationale des Sciences Appliquées
BP 77 Boulevard Beni Amir
25000 Khouribga
Morocco
yassine.maleh@ieee.org

ISBN 978-3-11-075533-6
e-ISBN (PDF) 978-3-11-076047-7
e-ISBN (EPUB) 978-3-11-076049-1
ISSN 2626-5443

Library of Congress Control Number: 2021950669

Bibliographic information published by the Deutsche Nationalbibliothek
The Deutsche Nationalbibliothek lists this publication in the Deutsche Nationalbibliografie; detailed bibliographic data are available on the Internet at http://dnb.dnb.de.

Cover image: wacomka/iStock/Getty Images Plus
Typesetting: Integra Software Services Pvt. Ltd.
Printing and binding: CPI books GmbH, Leck

MIX
Papier aus verantwortungsvollen Quellen
FSC® C083411

www.degruyter.com

About the editors

Mariyam Ouaissa (https://orcid.org/0000-0002-3993-8405) is a Ph.D. researcher associate in computer science and networks from Moulay Ismail University Meknes, Morocco, and a consultant trainer. She received her Ph.D. in 2019 from National Graduate School of Arts and Crafts, Meknes, Morocco, and her engineering degree in 2013 from the National School of Applied Sciences, Khouribga, Morocco. She is a communication and networking researcher and practitioner with industry and academic experience. Her research is multidisciplinary, which focuses on Internet of things, M2M, WSN, vehicular communications and cellular networks, security networks, congestion overload problem, and the resource allocation management and access control. She is serving as editorial board and reviewer for international journals and conferences, including *IEEE Access*, *Wireless Communications and Mobile Computing*, *International Journal of Smart Security Technologies* (*IJSST*), and *International Journal of Information Security and Privacy* (*IJISP*). Since 2020, she is a member of "International Association of Engineers (IAENG)" and "International Association of Online Engineering." Since 2021, she is an "ACM Professional Member." She has published more than 20 research papers (this includes book chapters, peer-reviewed journal articles, and peer-reviewed conference manuscripts), 5 edited books, and 5 special issues as guest editor. She has served on program committees and organizing committees of several conferences and events and has organized many symposiums/workshops/conferences as a general chair.

Zakaria Boulouard (https://orcid.org/0000-0002-4891-3760) is currently a professor in the Department of Computer Sciences, Faculty of Sciences and Techniques Mohammedia, Hassan II University, Casablanca, Morocco. In 2018, he joined the "Advanced Smart Systems" Research Team at the "Computer Sciences Laboratory of Mohammedia." He received his Ph.D. in 2018 from "Ibn Zohr University, Morocco" and his engineering degree in 2013 from the "National School of Applied Sciences, Khouribga, Morocco." His research interests include artificial intelligence, big data visualization and analytics, and optimization and competitive intelligence. Since 2017, he is a member of "Draa-Tafilalet Foundation of Experts and Researchers," and since 2020, he is an "ACM Professional Member." He has served on program committees and organizing committees of several conferences and events and has organized many symposiums/workshops/conferences as a general chair. He has served and continues to serve as a reviewer of numerous international conferences and journals. He has published several research papers, including book chapters, peer-reviewed journal articles, peer-reviewed conference manuscripts, edited books, and special issue journals.

Mariya Ouaissa (https://orcid.org/0000-0002-0088-3742) is a researcher associate and practitioner with industry and academic experience. She received her Ph.D. in 2019 in computer science and networks at the Laboratory of Modelisation of Mathematics, and in computer science from ENSAM-Moulay Ismail University, Meknes, Morocco. She is a networks and telecoms engineer, graduated in 2013 from National School of Applied Sciences Khouribga, Morocco. She is a co-founder and IT Consultant at IT Support and Consulting Center. She worked in School of Technology of Meknes Morocco as a visiting professor from 2013 to 2021. She is member of the International Association of Engineers and International Association of Online Engineering, and since 2021, she is an "ACM Professional Member." She is expert reviewer with Academic Exchange Information Centre (AEIC) and brand ambassador with Bentham Science. She has served and continues to serve on technical program and organizer committees of several conferences and events and has organized many symposiums/workshops/conferences as a general chair and also as editorial board and reviewer of numerous international journals. She has made her contributions in the fields of information security and privacy, Internet of things security, and wireless and constrained networks security. Her main research topics are Internet of things, machine-to-machine, device-to-device,

https://doi.org/10.1515/9783110760477-204

wireless sensor networks, cellular networks, and vehicular networks. She has published more than 20 research papers, 5 edited books, and 5 special issues as guest editor.

Yassine Maleh (http://orcid.org/0000-0003-4704-5364) is a cybersecurity professor and practitioner with industry and academic experience. He received his Ph.D. in computer science. Since 2019, he is working as a professor of cybersecurity at Sultan Moulay Slimane University, Morocco. He worked for the National Port agency (ANP) in Morocco as a senior security analyst from 2012 to 2019. He is a senior member of IEEE, member of the International Association of Engineers, and the Machine Intelligence Research Labs. He has made contributions in the fields of information security and privacy, Internet of things security, and wireless and constrained networks security. His research interests include information security and privacy, Internet of things, networks security, information system, and IT governance. He has published over 50 papers (book chapters, international journals, and conferences/workshops) and more than 10 edited and authored books. He is the editor in chief of the *International Journal of Smart Security Technologies*. He serves as an associate editor for *IEEE Access* (2019 Impact Factor 4.098), the *International Journal of Digital Crime and Forensics*, and the *International Journal of Information Security and Privacy*. He was also a guest editor of a special issue on *Recent Advances on Cyber Security and Privacy for Cloud-of-Things* of the *International Journal of Digital Crime and Forensics*, volume 10, issue 3, July–September 2019. He has served and continues to serve on executive and technical program committees and as a reviewer of numerous international conferences and journals such as *Elsevier Ad Hoc Networks*, *IEEE Network Magazine*, *IEEE Sensor Journal*, *ICT Express*, and *Springer Cluster Computing*. He was the publicity chair of BCCA 2019 and the general chair of the MLBDACP 19 symposium.

Mariyam Ouaissa, Mariya Ouaissa, Zakaria Boulouard, Sarah El Himer

Toward the Internet of vehicles: challenges and solutions

Abstract: Vehicle wireless networks are considered a key component of future intelligent transportation systems (ITS). They represent a promising technology which, when implemented, will offer the possibility of deploying a wide variety of applications. Some of these applications aim to improve road safety, and others aim to make the journeys of road users more comfortable. This new technology represents one of the most attractive areas of research in recent years, whether in academia or industry. To allow secure and reliable network access, vehicle-to-everything technologies are evolving to the Internet of vehicles (IoV) for increasing demands on emerging advanced vehicular applications. The IoV is based on the integration of new communication technologies, the integration of vehicular networks into the Internet of things, and optimal data processing. With these improvements, vehicular networks could support all ITS applications. In addition, new players could be interested in vehicular networks, attracted by the possible commercial applications in IoV. In this chapter, we have introduced various technological solutions that could be integrated into vehicular networks to meet the limitations of the vehicular architecture. Among these technologies, we can find software-defined networking, network function virtualization, edge computing, blockchain, big data, and artificial intelligence technologies. For each of these technologies, we present the concept of integrating them into the vehicle environment.

Keywords: ITS, IoV, V2X, LTE-V2X, 5G NR

1 Introduction

Intelligent transportation systems (ITS) describe all integrated applications as well as new communication and information processing technologies for the transport system. Thus, ITS make major contributions in several sectors of activity such as road

Mariyam Ouaissa, Moulay Ismail University, Meknes, Morocco,
e-mail: mariyam.ouaissa@edu.umi.ac.ma
Mariya Ouaissa, Moulay Ismail University, Meknes, Morocco,
e-mail: mariya.ouaissa@edu.umi.ac.ma
Zakaria Boulouard, LIM, Hassan II University, Morocco, e-mail: zakaria.boulouard@fstm.ac.ma
Sarah El Himer, Sidi Mohammed Ben Abdellah University, Fez 30000, Morocco,
e-mail: sarah.elhimer@usmba.ac.ma

https://doi.org/10.1515/9783110760477-001

safety or the development of surveillance systems. All of these new services aim to protect lives, to respect the environment as much as possible, and also to improve all factors related to driving, namely optimized navigation systems [1].

With the advent of wireless communication networks and the proliferation of mobile devices, the development of vehicular communication networks has emerged as a potential solution. Indeed, by establishing communications between vehicles, it becomes possible to guarantee an efficient transmission of information, concerning the condition of the road, the presence of an obstacle, or sudden braking. Vehicle communications must improve road safety, traffic flow, and the comfort of road users.

For the deployment of vehicular communication networks, a fully decentralized architecture, based on direct communications between vehicles, was first considered. This decentralized vehicular ad hoc network (VANET) was to ensure rapid transmission of information between neighboring vehicles, without requiring the deployment of expensive roadside infrastructure [2].

However, this decentralized architecture had different limitations. First, it only supported a single radio access technology, making the deployment of connected ITS applications more complex. Second, it only offered limited internet connectivity, which did not guarantee efficient and centralized data processing. Finally, it did not allow vehicles to interact with their environment (pedestrians, cyclists, surveillance cameras, etc.) [3].

In this context, vehicular networks have evolved toward a centralized, or hybrid architecture called Internet of vehicles (IoV). This architecture combines the benefits of an ad hoc approach and the benefits of a centralized approach. The IoV is based on three main ideas. Firstly, the integration of different communication technologies such as ITS-G5, LTE-V2X (vehicle-to-everything), and 5G NR (new radio). Then, vehicular networks are included in the Internet of things (IoT), allowing the establishment of communications between vehicles and surrounding objects. Finally, optimal data processing becomes possible, thanks to the permanent Internet connectivity offered by the IoV. With these developments, the IoV opens up vehicular networks to commercial applications, which is a major challenge for their adoption [4].

In this chapter, we offer an overview of the vehicle environment toward IoV. First, we begin by presenting the main characteristics, the main types of applications, and the main components defining vehicular networks. Then, we present the types of V2X communication and the different access technologies used in this system. Finally, we justify the evolution of vehicular networks from an ad hoc approach toward a centralized approach, namely, the IoV, and we introduce different technological solutions that could be integrated into the vehicular communication architecture to partially meet the limits of the ITS architecture.

2 Overview of vehicular networks

In this section, we present the main characteristics, main components, and main types of applications of vehicular networks.

2.1 Vehicular network characteristics

Vehicle networks, primarily based on direct communications between vehicles, have certain characteristics specific to mobile ad hoc network (MANET) communications. These include omnidirectional broadcast, variable rate wireless links, limited bandwidth, dynamic topology, short transmission range, variable node density, and limited physical security [5]. However, these networks also have certain characteristics linked to the fact that the nodes are vehicles:

- **Network topology:** due to the high-speed movement of vehicles, the network topology necessarily undergoes frequent modifications.
- **Frequent disconnections:** due to the high-speed movement of vehicles, breaks in communication links between vehicles or between vehicles and infrastructure are frequent.
- **Number of nodes:** due to the large number of vehicles in circulation, the vehicular networks must allow significant scaling.
- **Network density:** the number of vehicles in a given area depends on various factors such as the time, the geographical position, or the type of road.
- **Mobility models:** although the network topology is highly dynamic, vehicles generally follow the road layout and respective speed limits. Also, it is possible to define mobility models making it possible to reliably determine the future position of vehicles.
- **Bandwidth and latency:** many vehicular applications, in particular, are intended for road safety and require short transmission times. Without this, the proper functioning of these applications cannot be guaranteed. This in extreme cases could cause accidents. Likewise, transmission of a large amount of information must be possible when necessary (collision prevention, detection of an obstacle, etc.).
- **Energy and storage:** vehicles offer a quantity of energy, computing capacity, and storage space greater than those of common mobile objects. Also, although the development of low-energy-consuming solutions is a hot topic, vehicles support a greater number of operations and local processing.
- **The existence of areas of interest:** for many vehicle applications, the information generated is only relevant in a geographic area from ad hoc vehicular networks to the given IoV. This is particularly the case with information about the condition of the road, sudden braking, or cooperative data downloading. The existence of areas of interest is therefore an important characteristic of vehicular applications.

2.2 Components of vehicular communications

The primary idea of vehicular networks is to ensure road safety by rapidly transmitting information to the surrounding vehicles [6]. Also, two components have been specially designed to allow these communications (Figure 1):

- **On-board unit (OBU)**: on-board inside vehicles, this component allows vehicles to establish communication with surrounding terminal equipment and network equipment. The OBU is also used to process the information sent by vehicle sensors. OBU is a composed of different elements, in particular, sensors (GPS, LIDAR, etc.), a network interface (antenna) and a central control module.
- **Road side unit (RSU)**: this is a fixed wireless access point, positioned along roads and specifically designed for vehicular communications. The RSU has two essential roles. The first is to distribute the information locally, to vehicles directly connected to it and to neighboring RSUs. This RSU is also a gateway to the Internet network for the vehicles with which it is in communication.

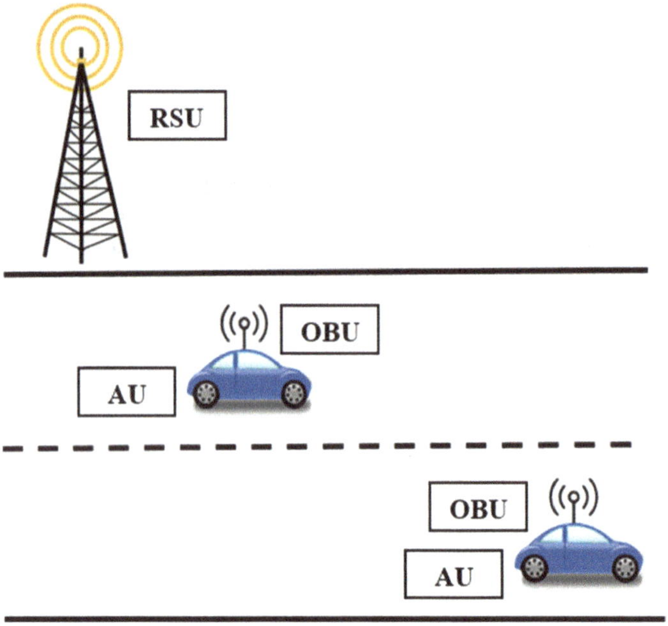

Figure 1: VANET architecture.

To these two components, we can add the application unit. It is an element integrated into the interior of vehicles and allowing the user to access the applications supported by this vehicle: road safety, entertainment, and so on. This component is necessarily coupled to the OBU to which it is connected by a wired or wireless connection. Indeed, the operation of the applications is based on the information carried by the OBU, the only means of communication between the car and its environment such as the Internet.

Considering the fact that vehicular networks must ensure the safety of all road users, we can add to the OBU and RSU elements not specific to vehicular networks. Base stations (BS), mobile terminals, connected traffic signs, and connected electric charging points could indeed prove useful in improving road safety and maximizing the amount of information available. This idea of integrating new components into vehicle networks to enhance road safety and to offer new services is one of the basic principles of IoV [7].

2.3 Type of applications

Many applications are envisaged for vehicular networks and are generally classified into three broad service categories: road safety, traffic flow, and passenger comfort and entertainment [8].

- **Road safety**: the primary purpose of ITS systems is to reduce the risk of road accidents. This can be achieved through systems by warning the driver of external risks and by setting up automatic reactions following the detection of these risks. Thus, for example, it is a question of detection of obstacles on the road, coordination of intersections, and cooperative collision prevention.
- **Traffic fluidity**: the primary purpose of these services is to manage the journeys of road users and to optimize them according to various factors: pollution, duration, cost, and so on. Thus, it is, for example, a question of managing traffic in real time, reserving parking spaces, or creating maps in a cooperative.
- **Vehicle maintenance**: the primary purpose of these services is to monitor the condition of the vehicle in real time in order to optimize maintenance costs and simplify its control. This involves, for example, remote maintenance and personal virtual assistant for simplified appointment scheduling.
- **Passenger entertainment**: the primary purpose of these services is to allow road users to be entertained and educated, and to provide visibility to advertisers. It is, for example, a question of multimedia applications, augmented reality, or points of interest.
- **Driver assistance**: the primary purpose of these services is to simplify driving the vehicle and limit the tasks assigned to the driver. It is, for example, a question of driving assistance, remote control of the vehicle, or total automation of the vehicle.
- **Health supervision**: the primary purpose of these services is to ensure well-being and health of the driver and his/her passengers. It is, for example, a question of cooperative detection of the level of fatigue, stress or blood alcohol level, adaptation of the environment to the mood of passengers, or medical assistance.

2.4 Requirements of vehicular network

Due to their number and diversity, vehicle applications necessarily have different needs. Various works have proposed an analysis of the performance criteria of the various vehicle applications. The main criteria are [9]:

- **Data rate** can be defined here as the number of bits transmitted per unit of time. It is generally expressed in bits per second. The minimum data rate can be calculated based on two parameters: the maximum number of messages transmitted per second and the size of a data unit.
- **Latency** can be defined here as the delay induced by the communication network for the transmission from the sender to the receiver of a data unit of an application. This latency can be measured at different levels of the Open Systems Interconnection (OSI) model. The maximum latency supported for the transmission of a message from the application layer of a sender to the application layer of a receiver will be called end-to-end latency.
- **Reliability** can be defined here as the probability that the value of a given parameter between layer n of a transmitter and layer n of a receiver is lower than the maximum value supported by the considered application. According to this definition, reliability can be measured at different levels of the OSI model. However, in the remainder of this document, we will focus on measuring reliability for end-to-end transmission, from the application layer of a transmitter to the application layer of a receiver.
- **The distribution area** can be defined here as the extent of the geographical area in which the data produced by an application must be distributed. A distribution area can characterize a set of applications.
- **The type of transmission** can be defined here as the transmission mode used for broadcasting the data generated by a given application. This type of transmission is defined here by three main parameters. The first of these is the nature of the messages: periodic or not. The second parameter is the broadcast mode: unicast, multicast, and broadcast. The third parameter corresponds to the types of communication used for data transmission: vehicle to vehicle (V2V), vehicle to pedestrian (V2P), vehicle to network (V2N), vehicle to infrastructure (V2I), and infrastructure to infrastructure (I2I).

3 Types of V2X communication

To allow the establishment of communications between the various components integrated into vehicular networks, the standardization of many types of communication, presented in Figure 2, is necessary [10]:

- **V2V:** this is the direct wireless communications between vehicles. These communications, firstly, allow the vehicle to locate itself in space, by retrieving various information from surrounding vehicles. These communications also allow the rapid dissemination of prevention messages: change of speed, change of direction, obstacle detection, and so on. These communications are aimed at reducing the risk of road accidents. V2V communications can rely on short-range communication technologies specifically designed for vehicular communications. They can also rely on cellular communications adapted to the vehicle environment.
- **V2I:** this represents the communication between a vehicle and equipment belonging to the roadside infrastructure: RSU, road lighting, traffic sign, radar, stop bus, and so on. These communications must first allow vehicles to better adapt to their environment. Indeed, the roadside equipment has different information: works, maximum speed change, and so on. In addition, this equipment aggregates the information sent by the vehicles: traffic density, detection of an obstacle, and so on. These communications also provide an Internet gateway to vehicles when RSUs are connected to the Internet. It should be noted that V2I communications are based on short-range communication technology designed for vehicular networks.
- **V2N:** this represents the communication between a vehicle and equipment belonging to the cellular communication network, particularly BS. These communications provide the functionality of V2I communications in areas not covered by roadside equipment RSU. Thus, BS can be used to transmit information between vehicles: condition of the road, accident, and other things. In addition, these V2N communications guarantee Internet access. Vehicles can therefore use them to connect to remote services. We can then talk about vehicle-to-cloud communications. These cloud services can in particular correspond to global road safety management or road traffic management applications.
- **V2P:** this represents the communication between a vehicle and a nonmotorized road user: pedestrian, cyclist, and others. These communications must make it possible to reduce the risk of collisions between vehicles and road users. To do this, V2P communications can rely on different technologies such as Bluetooth Low Energy, Ultra Wideband, and cellular network. These communications can also take different forms. First, unilateral; in this case, only one of the entities is notified of the arrival of the other entity (road user or vehicle). Then, bilateral; in this case, the two entities (road user and vehicle) are informed of the risk associated with the arrival of the other entity.
- **I2I:** these are wired or wireless communications between two network devices (BS and RSU). These communications firstly allow data to be distributed geographically. Thus, neighboring BS and RSU can exchange information (road safety, traffic, and entertainment) which will then be transmitted to the vehicles. These communications can also provide Internet connectivity to RSUs not directly connected to the network. In this way, it becomes possible to feed information back to cloud applications and to process more global data.

To these types of primary communication, we can add different types of communication currently considered: intelligent vehicle-to-grid and vehicle-to-building communications.

All types of vehicular communication are brought together under a global paradigm: the idea of V2X communication. V2X networks must allow the vehicle to communicate with any type of node existing in its environment. Thus, the proper functioning of road safety and traffic flow services could be guaranteed. In addition, new commercial applications, promoting the deployment of vehicular networks, could be designed.

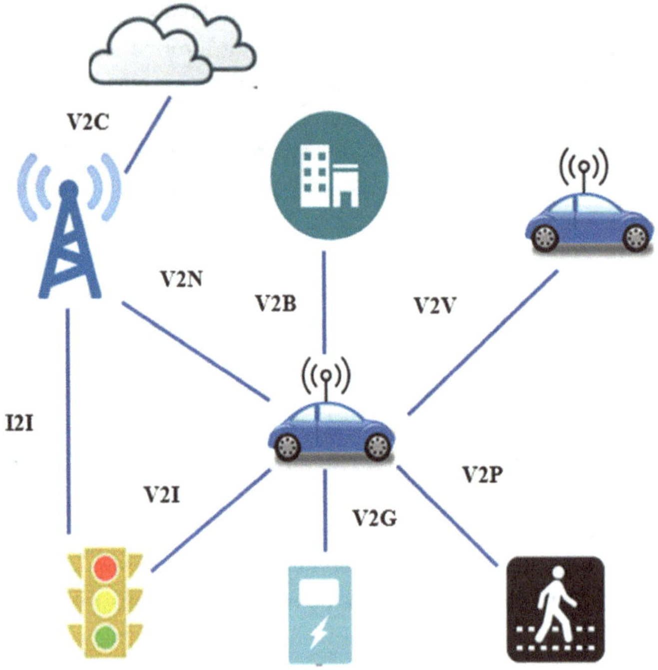

Figure 2: Types of V2X communication.

4 Access technologies in V2X communication

The types of communication V2V, V2I, V2N, even V2P, rely on two main types of access technologies: short-range communications, specially defined for vehicular networks, and cellular communication networks. This section aims to present these communication technologies. It should be noted that the integration of these access technologies into the vehicular communication architecture and their interoperability is provided for by the ITS reference architecture [11–12].

4.1 ETSI ITS-G5

The ITS-G5 is the solution offered by European Telecommunications Standards Institute (ETSI) for the deployment of high-performance vehicular communication networks. Today, cellular networks are gaining importance in the vehicle environment. However, the ITS-G5 was the first solution to be put forward by manufacturers and researchers for vehicular communications. This solution is based on the basic components of vehicular networks, RSU and OBU, and must make it possible to support V2V and V2I communications. The communication architecture of this ITS-G5 access technology, following the principle of the OSI model, is based on five main layers (Figure 3) [13]:

- **IEEE 802.11p:** derived from the IEEE 802.11a standard, IEEE 802.11p defines the physical and link layers of the OSI model for ITS-G5 technology. Originally created by the US Federal Telecommunications Commission, this standard offers improvements to allow the development of V2V and V2I communications. To guarantee resistance to the phenomena of attenuation and interference of the signal, and to multipath propagation, the IEEE 802.11p is based on a frequency band reduced by half compared to the IEEE 802.11a. This standard is also based on two main technologies, which are the orthogonal frequency division multiplexing technique, with an evolution of the guard interval and the transmission rate compared to the IEEE 802.11a, and the access method carrier sense multiple access with collision avoidance.
- **ITS-G5 decentralized congestion control (ITS-G5 DCC):** this decentralized congestion control in access layer is deployed within all ITS stations. This mechanism should make it possible to limit network congestion. To do this, it is based on an optimal allocation of resources between the ITS stations according to the level of occupation of the different communication channels. This is made possible by a precise measurement of the occupancy rate of the communication channels such as channel busy ratio.
- **ITS geo networking (GeoNet) and construction:** the idea of geographic routing was originally proposed for mobile wireless networks. It then appeared as a relevant solution for geographically distributing the data generated by ITS applications. The operation of this protocol is based on the regular exchange of messages between vehicles and on the definition of the location table at the level of each vehicle. By using these two mechanisms, a distribution of data, in a given geographical area, becomes possible. The basic transport protocol (BTP), similar to the UDP protocol, provides a solution for transporting the GeoNet protocol, as well as ITS messages from the upper layers.
- **ITS messages:** these C-ITS messages correspond to the definition of standardized functions necessary for the operation of ITS applications. Two main categories of messages are specified for road safety: cooperative awareness messages (CAM) and messages decentralized environmental notifications message (DENM). CAM messages should allow vehicles to better adapt to their environment through a

regular exchange of information regarding their status and their attributes. The messages DENM must allow a rapid reaction to any abnormal event through effective dissemination of information to vehicles located in the area affected by this event.

- **The applications:** the effectiveness of these applications, intended for road safety, traffic flow, or the entertainment of road users, is based on efficient operation of the protocol stack that has just been presented: IEEE 802.11p, ITS -G5 DCC, GeoNet, BTP, and CAMs/DENMs.

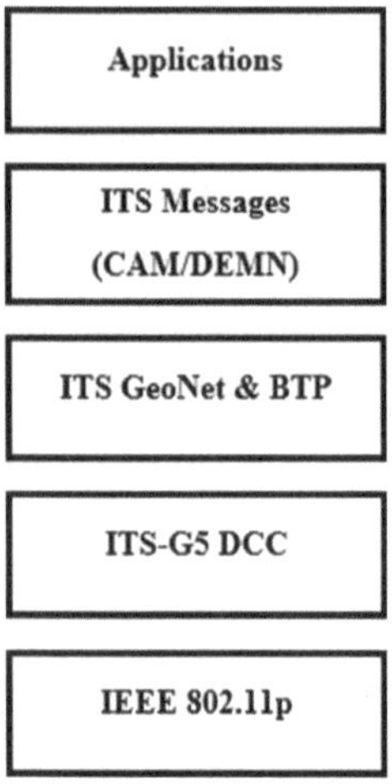

Figure 3: ITS-5G protocol layers.

4.2 Cellular networks

Cellular networks are the other approach considered for the deployment of vehicular communication networks. These cellular networks have an important advantage over ITS-G5 networks: an existing large-scale deployment guaranteeing good coverage. They are also able to provide decent data throughput and latency. Two generations of cellular communication networks could allow the deployment of high-performance vehicular communications: 4G networks (LTE-V2X) and 5G networks. These two generations of cellular networks are quickly introduced in this section [14].

4.2.1 LTE-V2X

The idea of integrating vehicular communications into cellular networks was first proposed in 2015 in Release 14 published by the third-generation partnership project (3GPP), the body behind the specification of 3G and 4G cellular networks. Different names are used to name these cellular networks intended for vehicular communications: LTE-V, LTE-V2X, or even cellular-V2X. This integration of vehicular communications, requiring optimal use of communication channels and high reliability, is based on two major ideas:

- **A new radio interface:** cellular communications generally rely on the use of a single radio interface, the LTE-Uu interface. This interface allows a mobile terminal to establish communication with a base station. Also, to allow direct communications between vehicles, it was necessary to define a new interface: the PC5 interface. This interface presents various improvements to ensure its operation in the vehicle environment: reduction of latency, improvement of the scaling up, and consideration of high mobility. It should be noted that for this PC5 interface, we speak of a secondary as opposed to the primary communications between vehicles and base stations.
- **Object-to-object communication technology:** cellular communications normally rely on communications between vehicles and base stations (V2N). However, vehicular networks require direct communications between vehicles (V2V) to be able to guarantee low latency and high road safety. To deliver these V2V communications in the cellular environment, 3GPP is considered using an existing technology named Prose. With this technology, LTE cellular networks could support the establishment of direct object-to-object short-range communications. Indeed, thanks to the use of new communication channels, Prose could allow vehicles to discover and establish direct communication with their neighbors, without going through the base station.

4.2.2 5G NR

5G NR is a new radio access technology developed by 3GPP. It should serve as the basis for fifth-generation cellular networks. These fifth-generation cellular networks should support three main types of communication:

- Enhanced mobile broadband: to allow the management of multimedia applications requiring very high data rates.
- Ultra-reliable and low-latency communications (URLLC): to allow the management of communications with significant constraints in latency, reliability, and availability. The URLLC is intended in particular for vehicle applications.
- Massive machine-type communication: to allow the management of a large number of direct communications between machines, in particular for the IoT. To provide these services, 5G NR radio access technology requires various improvements over existing technologies: frequency spectrum management, latency, scaling, and reliability. To do this, the integration of different techniques is considered. We speak in particular of full duplex in frequency band, millimeter waves, nonorthogonal multiple access, and massive multiple input–multiple outputs.

With these developments, 5G NR technology could, like LTE-V2X technology, support vehicular communications, while guaranteeing much higher performance (throughput and latency). Also, this standard, which should be deployed on a large scale in the years to come, today appears to be the future of cellular communication networks and vehicular communication networks. It should be noted that two deployment modes are considered for the deployment of 5G NR technology: nonautonomous mode (non-standalone) and autonomous mode (standalone). Nonautonomous mode is a case where 5G NR technology relies on LTE communications infrastructure for signaling control. 5G NR technology then focuses only on the transfer of information. Autonomous mode corresponds to a case where 5G NR technology is based on a 5G infrastructure and core network. The 5G NR technology then manages the transfer of information and the control of signaling. It can also be noted that 5G networks are designed to allow the simultaneous use of several radio access technologies such as multiple radio access technologies for the same mobile terminal. It would then become possible for a vehicle to receive data from other vehicles while transmitting, at the same time, data to a base station or to other vehicles. This approach offers important prospects in terms of capacity and throughput for cellular and vehicular networks.

5 Technological solutions for Internet of vehicle communication

The IoV must make it possible to meet the needs of connected ITS applications. To do this, the IoV is based on three principles: the integration of new communication technologies, the integration of vehicular networks into the IoT, and more efficient data processing.

However, these various developments only provide a partial response to the current limits of vehicular networks. Indeed, various points must still be considered so that the communication architecture can support all ITS applications. These necessary improvements can be classified into different categories, corresponding to the different plans of the ITS reference architecture, namely control and data, rapid network reconfiguration, optimal management of resources, interoperability of heterogeneous networks, for example, the integration of heterogeneous access networks (5G NR, LTE V2X, and ITS-G5), acceleration of the application life cycle, flexible deployment, security, and respect for privacy [15].

Faced with these evolutionary needs, many improvements to vehicular architecture have already been proposed in the literature. This part is generally based on the use of innovative technological solutions.

Different technological solutions could be integrated into the vehicular communication architecture: software-defined networking (SDN), network function virtualization (NFV), edge computing (EC), blockchain (BC), big data, and artificial intelligence.

Each of these technologies could make it possible to partially respond to the limits of the vehicular communication architecture [16].

Some limitations of the ITS reference architecture are related to the technologies on which this architecture is based. This is especially true for the lack of flexibility, the lack of dynamicity, and the lack of upscaling. To overcome these limitations, it would therefore seem relevant to consider other technological solutions. This is why, in this section, we present some of the technological solutions that could be considered.

5.1 Software-defined networking (SDN)

In traditional communication networks, routers are responsible for two main operations, the selection of the communication path and the transfer of data. We thus speak of control plane and data plane. Usually, the control plane relies on local decision making and distributed routing protocols such as Open Shortest Path First or Border Gateway Protocol. These protocols allow each router to retrieve information about the state of the network and determine how the information should be routed [17].

However, this approach has an important limitation: the lack of flexibility. Indeed, the distributed control functions are based on routing rules preinstalled when the equipment is designed. Changing these rules would require costly and complex modification of all routers. Also, quickly developing a large-scale network seems unlikely with this approach. SDN technology could help overcome these limitations. Indeed, it aims to guarantee greater dynamicity, rapid changes in network control policies, and a reduction in operating expenses. To do this, this technology is based on a major idea: a decoupling of the control plane and the data plane.

Thus, with SDN, routers remain in charge of transferring data but are no longer responsible for selecting the optimal path. This decision making is centralized with SDN and based on knowledge of the entire network. This should in particular make it possible to guarantee better decision making, improving the overall network performance.

In the SDN approach, centralized decision making is carried out by a new element, integrated into the architecture: the SDN controller. This controller is connected to all network equipment and is responsible for managing their routing tables. To do this, these elements send the SDN controller different types of information that will allow it to perform network functions: routing, load balancing, and so on.

Depending on the extent of the communication network, and the number of nodes to be managed, the SDN control plane can be considered centralized or logically or even physically distributed. In this second scenario, several SDN controllers are used. Each of these controllers manages a given area of the network and has a global view of the state of the network through exchanges with other SDN controllers.

SDN, decoupling the control plane and the data plane, represents an important evolution in the way of thinking about communication networks. With SDN, the routing table is replaced at the level of each router by a flow table. This is directly administered by the SDN controller and is used by the network device to manage the data transfer.

5.2 Network function virtualization (NFV)

The use of network functions currently relies on the deployment of dedicated proprietary hardware equipment. Establishing service chains, or increasing the capacity of an existing service chain, therefore, requires the purchase and maintenance of a set of expensive, proprietary hardware equipment.

NFV technology, supported by the ETSI and the Open Networking Foundation, aims to solve this problem. Indeed, NFV must make it possible to offer a software version of the network functions and thus to dissociate network functions and hardware proprietary equipment.

This technology therefore constitutes an important evolution in the way of designing, deploying, and managing network services. With NFV, it becomes possible to run a router, firewall, or encryption service in a virtual machine deployed on a standard X86 server and not on proprietary hardware.

The advantages of this approach are numerous. First of all, we can talk about a significant reduction in costs, both purchasing and maintenance. We can also talk about increased flexibility through flexible deployment of virtual machines.

Finally, associated with this idea of flexibility, another important advantage of the software approach is the possibility of rapidly deploying new services or rapidly improving existing services.

5.3 Edge computing (EC)

The development of new services is today based on an essential paradigm cloud computing. This approach consists of using remote servers, public or private, made available by service providers, to store and process data. The user is then no longer responsible for deploying and maintaining the infrastructure (infrastructure as a service), the platform (platform as a service), or even the application he/she wishes to use (software as a service). Thus, the cloud approach simplifies the storage and processing of data from the user's point of view. It also allows cost reduction, great flexibility, high bandwidth, and high reliability. This is why the development of ITS applications based on a cloud approach has been proposed in many works in literature.

However, this cloud approach has various limitations. Among these limitations, we can cite the latency resulting from the centralization of data processing. This latency could cause road safety or traffic flow problems for applications requiring high reactivity. Another blocking point is the move to scale. Indeed, with a constantly increasing number of vehicles, the feedback of unprocessed and unfiltered information could lead to network congestion. A third limitation is the Internet connection necessary for the operation of this system. Without this connection, access to the service becomes impossible. In the vehicular environment, a vehicle not necessarily being permanently connected, this could also constitute a significant brake. A final potential limitation is the lack of adaptation to the context. For many ITS applications, this adaptation is necessary and therefore must be enabled by the communication architecture.

Therefore, to overcome these various limitations, a new paradigm has been introduced: EC. EC extends the cloud approach to the edges of the network. The main idea of EC is therefore to offer outsourced services at the edge of the network and not in remote data centers. To do this, the EC uses the communication, storage, and computing capacities of all network equipment: routers, base stations, vehicles, and other things.

This technology therefore constitutes an important evolution in the way of thinking about outsourced services. Unlike the cloud approach, with EC, operations are carried out as close as possible to the user. The deployment of flexible, reliable, and secure services is therefore guaranteed. In addition, the EC responds to the limitations of the cloud approach: latency, network congestion, need for a permanent Internet connection, and lack of consideration of the context.

It can be noted that this paradigm is carried by different actors. Firstly, the mobile operators, led by ETSI, introduced the idea of mobile edge computing or multi-access edge computing (MEC). Secondly, the network equipment manufacturers, led by Cisco, introduced the idea of fog computing. Finally, the service providers, led by Google, introduced the idea of cloudlet.

These different implementations (MEC, fog computing, and cloudlet) have the same objective, that is, to deploy services at the edge of the network. However, these implementations are based on different equipment and different virtualization layers. In addition, each of these approaches has specific benefits, namely, the adaptation to the context, bandwidth, latency, and cost of deployment [18].

5.4 Blockchain (BC)

The implementation of security and privacy protection services is generally based on a centralized solution, called public key infrastructure (PKI). This approach is based on a simple idea. A central trusted authority guarantees the reliability of information relating to all network users. This approach, with proven operation and reliability, has been adopted by various pilot projects for the deployment of ITS, like the SCOOP and PRESERVE projects.

However, PKIs have limitations. The first of these is the existence of a central certification authority (CA), a major and single point of failure in the network. Compromise of this element could result in loss of security for the entire network. Another important limitation is the lack of reactivity, in particular, for the management of certificate revocation. Thus, a malicious user could continue to communicate at the expense of security. A third limitation is the lack of transparency in the management of revocation lists. These are not immutable, so a CA can change them without the knowledge of users. In order to provide an answer to these various limitations, a new, decentralized solution has been proposed: the BC. This technology, originally designed for the secure exchange of cryptocurrencies, has been extended to other areas, including vehicular networks. The BC makes it possible to store and transmit information without a central control body.

Thus, with the BC, a set of nodes, not dependent on a single entity, can guarantee the integrity of a distributed ledger. This integrity is enabled through the use of consensus mechanisms and cryptographic techniques.

BC is an important evolution in the way of thinking about network security. Indeed, as opposed to an approach based on a centralized PKI, the BC could allow the deployment of decentralized PKIs. This architectural evolution could make it possible to limit the risk of attacks linked to the existence of a single point of attack. With each node acting independently, the attacker should be able to take control of the majority of the nodes in order to influence the behavior of the network. In addition, this technology, designed to ensure rapid dissemination of information between the different nodes of the network, could allow more efficient certificate revocation. Finally, by making the information available network nodes and by ensuring data immutability, this technology offers a greater level of transparency [19].

5.5 Big data and artificial intelligence

Current communication networks generate significant amounts of information: geolocation, duration of calls, use of applications, and so on. Efficient use of this information could allow operators to optimize the costs and operation of existing services: deployment, maintenance, resource allocation, security, routing, and so on. New income could also be generated. Also, the processing of this information appeared to be a major axis of development.

However, storing, accessing, and processing such large volumes of data are not possible with traditional methods of storage and access. It is for this reason that new tools for processing massive data (big data), in particular, carried by service providers (Google, Microsoft, etc.) have emerged. These tools enable the collection of data from multiple sources: the aggregation and processing of massive volumes of data. Apache Hadoop and Apache Spark are among the most widely used big data tools.

However, the added value of these platforms does not lie solely in the storage and access to this data. In fact, it is the processing that can be carried out on this data that will allow decision making beneficial to the management of the network. To do this, artificial intelligence (AI) techniques are now a highly acclaimed approach. Indeed, thanks to the development of equipment with significant computing capacities, the application of AI techniques to large volumes of data has become possible. Thus, reliable prediction models can be developed and implemented.

By coupling efficient collection and processing of big data with powerful prediction models, management, control, and security could be improved. Also, the application of these technologies to the vehicle environment, subject to significant variations, seems relevant [20].

6 Conclusion

In recent years, the automotive world has increasingly relied on wireless communication systems in order to meet safety requirements. Accident alerts, the identification of potentially dangerous obstacles, and the sharing of information on traffic conditions are services considered fundamental to improving driving and ensuring road safety. In order to make these services possible, a new concept dedicated to the development of transport systems has emerged, called ITS. In this chapter, we have offered an overview of the vehicle environment. To do this, we started by presenting the main characteristics: the main types of applications and the main components defining vehicular networks. Next, we described the different types of V2X communication as well as main types of access technologies. Next, we justified the evolution of vehicular networks from an ad hoc approach toward a centralized approach, namely, the IoV. Considering that the IoV represents the future of vehicular networks, we identified the current limits of the IoV and introduced the various technological solutions that could be integrated into vehicular networks to meet the limits of the vehicular architecture. For each of these technologies, we presented the concept and the benefits that their integration into the vehicle environment could present.

References

[1] Dimitrakopoulos G, Demestichas P. Intelligent transportation systems. *IEEE Vehicular Technology Magazine*, 5(1):2010.

[2] Ouaissa M, Ouaissa M, Houmer M. *Secure Hierarchical Infrastructure-Based Privacy Preservation Authentication Scheme in Vehicular Ad Hoc Networks*. CRC Press, USA, 2021.

[3] Houmer M, Ouaissa M, Ouaissa M, Hasnaoui ML. Secured and enhanced greedy perimeter stateless routing protocol for vehicular ad hoc networks. *International Journal of Interactive Mobile Technologies (iJIM)*, 14(13):2020.

[4] Sharma S, Kaushik B. A survey on internet of vehicles: Applications, security issues & solutions. *Vehicular Communications*, 20:2019.

[5] Cunha F, Villas L, Boukerche A, Maia G, Viana A, Mini RA, Loureiro AA. Data communication in VANETs: Protocols, applications and challenges. *Ad Hoc Networks*, 44:2016.

[6] Ouaissa M, Ouaissa M, Houmer M, Rhattoy A. *Improved MAC Design-based Dynamic Duty Cycle for Vehicular Communications over M2M System*. Springer, Singapore, 2021.
[7] Ouaissa M, Houmer M, Ouaissa M An enhanced authentication protocol based group for vehicular communications over 5G networks. In *2020 3rd International Conference on Advanced Communication Technologies and Networking (CommNet)*, IEEE, 2020.
[8] El Faouzi N, Leung H, Kurian A. Data fusion in intelligent transportation systems: Progress and challenges–A survey. *Information Fusion*, 12(1):2011.
[9] Mohammad SA, Rasheed A, Qayyum A. VANET architectures and protocol stacks: A survey. In *International Workshop on Communication Technologies for Vehicles*. Springer, Berlin, Heidelberg, 2011.
[10] Jgian W, Yameng S, Yuming G, Rundong Y. A survey of vehicle to everything (V2X) testing. *Sensors*, 19(2):2019.
[11] Yoshizawa T, Preneel B Survey of security aspect of V2X standards and related issues. In *2019 IEEE Conference on Standards for Communications and Networking (CSCN)*. IEEE, 2019.
[12] Mannoni V, Berg V, Sesia S, Perraud E A comparison of the V2X communication systems: ITS-G5 and C-V2X. In *2019 IEEE 89th Vehicular Technology Conference (VTC2019-Spring)*. IEEE, 2019.
[13] Tahir MN, Katz M. Heterogeneous (ITS-G5 and 5G) vehicular pilot road weather service platform in a realistic operational environment. *Sensors*, 21(5):2021, 2021.
[14] Abdel Hakeem SA, Hady AA, Kim HW. Current and future developments to improve 5G-NewRadio performance in vehicle-to-everything communications. *Telecommunication Systems*, 75(3):2020.
[15] Abdel Hakeem SA, Hady AA, Kim HW. 5G-V2X: Standardization, architecture, use cases, network-slicing, and edge-computing. *Wireless Networks*, 26(8):2020.
[16] Fadhil JA, Sarhan QI Internet of vehicles (IoV): A survey of challenges and solutions. In *2020 21st International Arab Conference on Information Technology (ACIT)*. IEEE, 2020.
[17] Abbas MT, Muhammad A, Song WC. SD-IoV: SDN enabled routing for internet of vehicles in road-aware approach. *Journal of Ambient Intelligence and Humanized Computing*, 11(3):2020.
[18] Hammoud A, Sami H, Mourad A, Otrok H, Mizouni R, Bentahar J. AI, blockchain, and vehicular edge computing for smart and secure IoV: Challenges and directions. *IEEE Internet of Things Magazine*, 3(2):2020.
[19] Labrador M, Hou W Implementing blockchain technology in the internet of Vehicle (IoV). In *2019 International Conference on Intelligent Computing and its Emerging Applications (ICEA)*, IEEE, 2019.
[20] Xu W, Zhou H, Cheng N, Lyu F, Shi W, Chen J, Shen X. Internet of vehicles in big data era. *IEEE/CAA Journal of Automatica Sinica*, 5(1):2017.

Inam Ullah Khan, Muhammad Abul Hassan,
Asrin Abdollahi, Bisma Baig

AI-enabled communication protocols for vehicular ad hoc networks (VANETs): fundamentals and future challenges

Abstract: Vehicular ad hoc networks (VANETs) merging with 5G technology will result in cognitive transportation system. As for optimal data transmission, 5G is utilized to reduce the burden on mobile networks. VANETs are emerged from mobile ad hoc networks. Parameters like distance, nodes movement, mobility models, link stability, and speed need to be improved accordingly. This survey study mainly focuses on the types of ad hoc networks, comparative analysis of wireless communication technologies, and routing protocols for VANETs. Also, different techniques like machine learning, deep learning, metaheuristic search algorithms, and optimization techniques are discussed to enhance the capabilities of VANETs. In addition, for secure communication in VANETs, intrusion detection systems are studied to identify different security attacks.

Keywords: VANETs, IoT, 5G, AODV and routing

1 Introduction

Vehicular transportation is considered the most intelligent technology. The transport industry needs better communication channels in between vehicles. The main aim of vehicular ad hoc networks (VANETs) is to provide safe environment to mankind. VANETs are largely used to avoid accidents and help humans in decision making. Earlier, there were problems in traffic jams, and when vehicle changes the tower from one place to another, end-to-end delay is considered the main problem. Mostly, people send data while they are traveling, so VANETs provide optimal communication links. Ad hoc networks are considered having no fixed network structure. VANETs are evolved from ML. In mobile ad hoc networks (MANETs), the nodes

Inam Ullah Khan, Department of Electronic Engineering, School of Engineering and Applied Sciences, Isra University, Islamabad, Pakistan, e-mail: inamullahkhan05@gmail.com
Muhammad Abul Hassan, Department of Computing and Technology, Abasyn University, Peshawar 25000, Pakistan, e-mail: abulhassan900@gmail.com
Asrin Abdollahi, Department of Electrical Engineering, University of Kurdistan, Sanandaj, Iran, e-mail: asrin.abdollahi2961@gmail.com
Bisma Baig, Department of Software Engineering/COMSATS University, Abbottabad Campus, Pakistan, e-mail: bismabaig007@outlook.com

https://doi.org/10.1515/9783110760477-002

are static, while in VANETs the nodes move on ground. Due to the dynamic nature of VANETs, deployment of routing techniques are so difficult. The architecture involves vehicle-to-vehicle communication and vehicle-to-infrastructure communication. Also, the vehicles use fixed access points as well. In addition, hybrid architecture merges with ad hoc communication in VANETs [1].

VANETs can be utilized in many applications which include improvement in passenger comfort, safety, emergency warnings, road bad condition, and signal warning. Apart from that, wireless communication technology is considered the backbone. IEEE standards can be used for short-range communication, where the maximum range of applications is about 1,000 m in normal highway. VANETs information data is divided into two types: safety and comfort messages. However, in VANET, the topology is changing rapidly which directly causes connectivity issues. MAC layer is used as a standard for message dissemination, clustering, and also for power assignments [2].

Vehicular ad hoc routing is used to support unicast communication within nodes. Each routing protocol must support IP connectivity. However, handshake process is not appropriate while using routing protocols in communication networks. VANETs need to consider optimal routing, security, and available mechanisms to check the performance. Cryptography techniques can be utilized to secure communication in between vehicular nodes. Trustworthy keys need to be established to use the piggyback method using TROPHY messages. TAD-HOC environment is used to perform the simulation of different variants of AODV routing protocol. The proposed protocol is compared with the existing scheme which shows better results. I-AODV, AODV-R, and AODV-L are utilized in the simulation [3].

VANETs consist of wireless devices like GPS for positioning, where nodes are connected with the base station. Also, WiMax and cellular networks are considered the prominent technology in VANETs. Vehicular networks consist of node mobility, reliability, and predictability of mobility using road directions for traffic management. WLANS and WPANS are used for multihop communication in dynamic environment [4].

Internet of vehicles is used as a special class of VANETs. This area is the combination of VANETs and Internet of things (IoT). The basic framework of vehicles uses to interact with the IoT sensor nodes which take data packets from traffic, road side units, and also handheld devices. VANETs are having so many problems such as congestion, end-to-end delay, security, and optimization-related issues [5]. Figure 1 illustrates the basic concept of 5G technology with VANETs. Inter-roadside, vehicle-to-roadside, and vehicle-to-vehicle communication can be efficiently made possible utilizing 5G networks.

The major points of this survey are as follows:

- Ad hoc networks are MANET, VANET, and SANET
- Routing protocols for VANETs
- Proactive, reactive, and hybrid routing protocols
- Wireless communication technology for VANETs
- Comparison of 1G to 6G

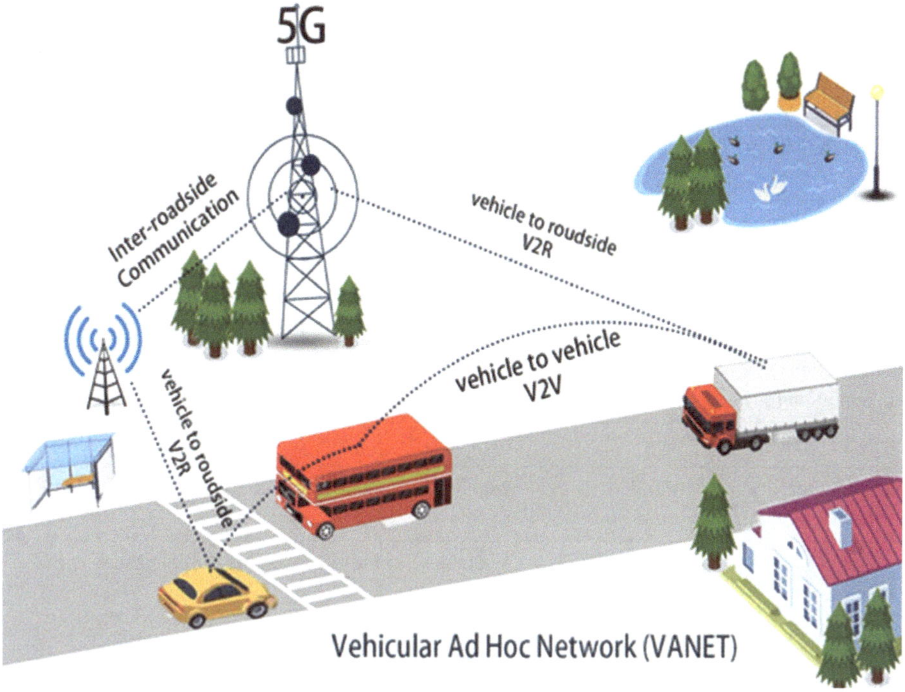

Figure 1: Vehicular ad hoc network using 5G technology.

2 Literature survey

The demand of routing in the field of ad hoc networks has changed the dynamics. Selection of optimal path in networks needs to improve quality of service, user experience, throughput, and end-to-end delay. Metaheuristic search algorithms like ant colony optimization help communications to improve network performance [12]. From base station to nodes, wireless communication is used to enhance decision tree algorithm [13]. 3D centroid algorithm and path loss model can be used in wireless communication to strengthen the received signal strength. RSSI is the possible technique to check the signal from base station to nodes [14]. However, energy-efficient routing improves the lifetime of ad hoc networks [15]. Ad hoc networks use traditional routing protocols, including AODV, TORA, DSDV, OLSR, and FSR [16]. Table 1 describes the idea of ad hoc networks like MANETs, VANETs, and flying ad hoc networks (FANETs).

Table 1: Ad hoc networks.

Performance metrics	Mobile ad hoc networks	Vehicular ad hoc networks	Flying ad hoc networks
Network topology change	Low	Medium	Very fast
Radio signal propagation	Close to ground (signal)	Close to ground (signal)	Not close to ground (signal)
Node's energy consumption	Low	Low	Not high for mini drones and high for large drones
Computation consumption (power)	Low	High	High
Node's mobility	Low	High	Very High
Network mobility model	Random selection	Regular selection	Regular under certain condition
LoS	Present in all cases	Available in some cases	Available in some cases
Security	High	High	High
Node's localization	GPS	GPS, AGPS	IMU, GPS, DGPS
Node's density	Low	High	Low
Update mechanism	Not often	Often	Often
Simulator	Ns-2, NS-3, OPNET	OMNET++, NS-2, NS-3, Matlab	NS-2, NS-3, Matlab, QualNet

3 Wireless communication technologies for VANETs

Wireless communication technologies are having an important role in VANETs. Therefore, from 1G to 6G, data capability, multiplexing, switching, and different frequencies are used. Table 2 describes the comparison of different wireless technologies. Apart from that, these communication networks help enhance different metrics in VANETs. In early 1970s, Bluetooth technology was used in communication networks. The capacity of 1G was hardly 2 kB/s and FDMA. Then this technology get advanced with the passage of time, and LTE was used for the first time in 2010. Therefore, wireless communication technology gets evolved, where 2020 satellite network theory is utilized for dynamic network scenarios.

While 5G can be utilized in VANET-based intelligent traffic management and accident prevention, this network is also used to reduce delay. 5G with VANETs are

Table 2: Comparative analysis of wireless technologies.

Generations (G)	Appeared in	Capacity of data	Technology used	Multiplexing	Network switching	Used in	Frequency
First generation (1G) [6]	1970–1984	2 kB/s	Analog system	FDMA	Circuit switched	Voice over-only	800–900 MHzz
Second generation (2G) [7]	1990–2003	10–473 kB/s	Digital GPRS, EDGE	TDMA and CDMA	Packet switched	Voice, data, MMS	850–1,900 MHz
Third generation (3G) [8]	2000–2003	384 kB/s to 30 MB/s	GSM, 3 GPP	CDMA	Packet switched	High-speed Internet	1.6–2.5 GHz
Fourth generation (4G) [9]	2010	200 MB/s to 1 GB/s	LTE	MC-CDMA	Packet switched	VOLTE, video calling support	2–8 GHz
Fifth generation (5G) [10]	2015	>1 GB/s	IPV6	CDMA	Both packets (switched and packet)	Dynamic information access	>4 GHz
Sixth generation (6G) [11]	2020	9 GB/s	OWE	Satellite network theory	Both packets (switched and packet)	Dynamic information access	100 GHz to 3 THz

illustrated in Figure 2. Ground-based vehicles are moving on the road, which are connected with the base station using 5G network.

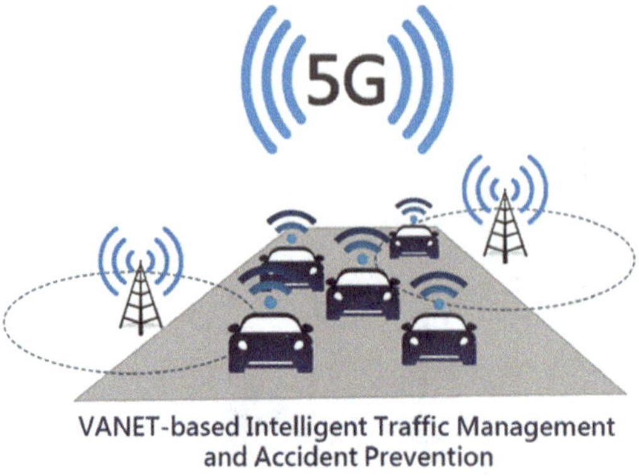

Figure 2: VANET-based intelligent traffic management using 5G networks.

4 Ad hoc networks

Ad hoc networks are having no infrastructure. Figure 3 describes three categories of ad hoc networks. Therefore, ad hoc networks are divided into three main categories which are mentioned as follows.

MANETs: These are the collection of static nodes which can be located dynamically. The nodes are continuously changeable due to the ever-changing topological structure [18].

VANETs: These are the combination of ground-based nodes which are moving and connected with the base station using wireless networks.

FANETs: These are directly referred to the UAV network [19].

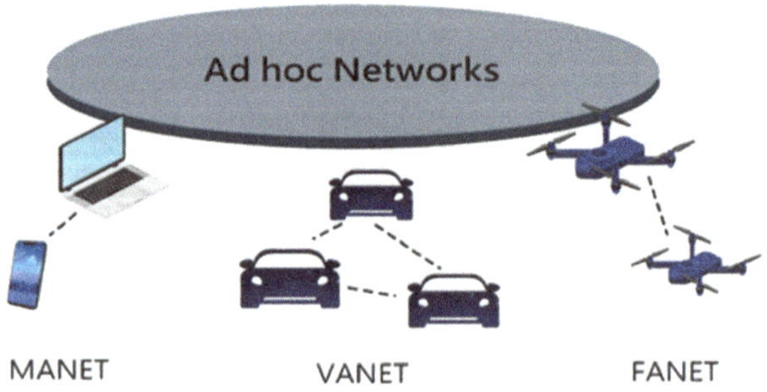

Figure 3: Ad hoc networks (MANET, VANET, and FANET).

5 Cloud computing in vehicular ad hoc networks

VANETs use to store a lot of data packets using cloud computing [20]. VANET-based cloud computing helps improve different transport applications which use to ensure safety, data aggregation, security, and resource management [21]. Figure 4 represents the hybrid scenario which involves UAVs and ground-based vehicles which try to store possible information in the cloud and send the data packets to the server.

6 Routing protocols for VANETs

The increasing demand in wireless communication can be utilized in VANETs. The vehicles are mostly based on metrics that improve traffic management. Therefore, routing protocol enhances VANET channels. In VANETs, each node acts like a router which can transmit information between vehicles. Routing protocols are divided in below mentioned categories [22]. Figure 5 shows topology-based routing protocols which include proactive, reactive, and hybrid.

Proactive routing
In this approach, routing tables receive updated information which regularly exchanges information between vehicular nodes [23].

Reactive routing
Reactive routing is quite a changed technique in comparison with proactive, where routing tables are updated on demand [23].

Hybrid routing
Hybrid routing is the combination of both proactive and reactive routing [23].

Routing protocols based on position/location
In these protocols, the position or location needs to be send in data packets from source to destination [24].

Cluster-based routing protocols
Cluster-based routing is usually defined by identifying leader or cluster head in each network [23]. Figure 6 shows the concept of clustering in VANETs.

Energy-efficient routing
These routing protocols only focus on energy efficiency which improve the quality of experience metrics. Figure 7 shows the different categories of VANET-based routing techniques.

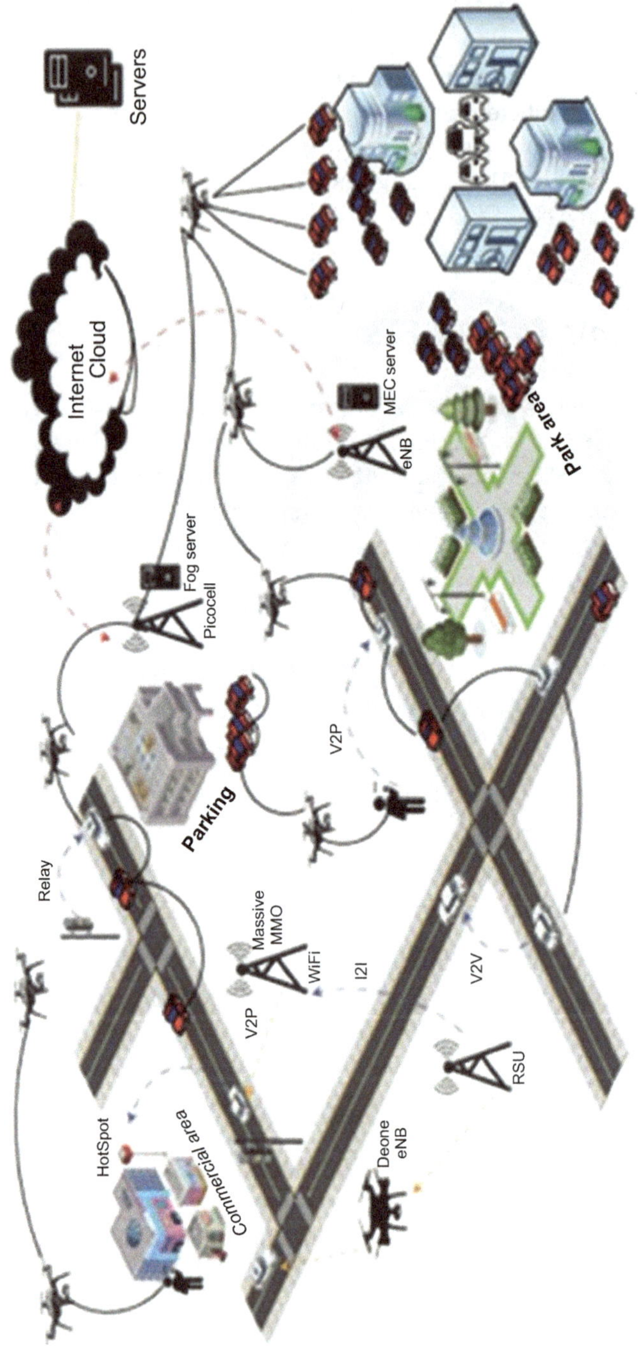

Figure 4: UAV-based VANETs using cloud computing.

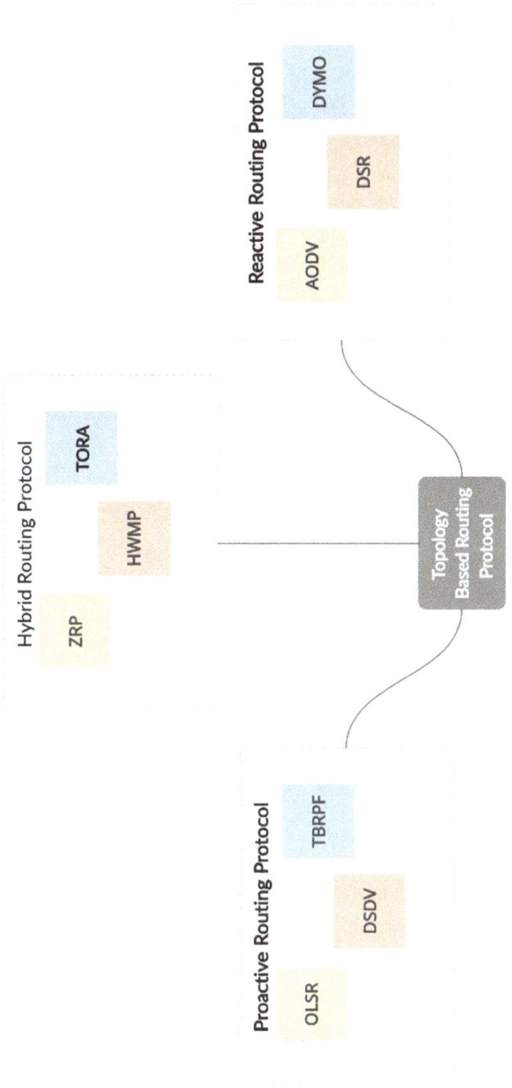

Figure 5: Topology-based routing protocols.

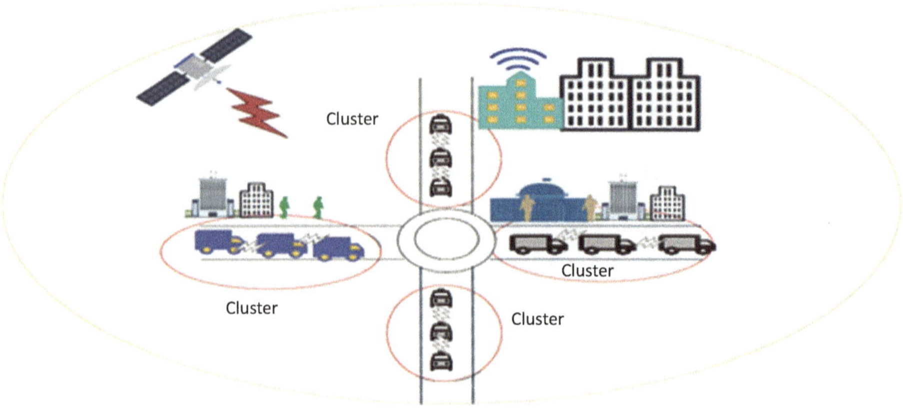

Figure 6: Clustering in VANETs.

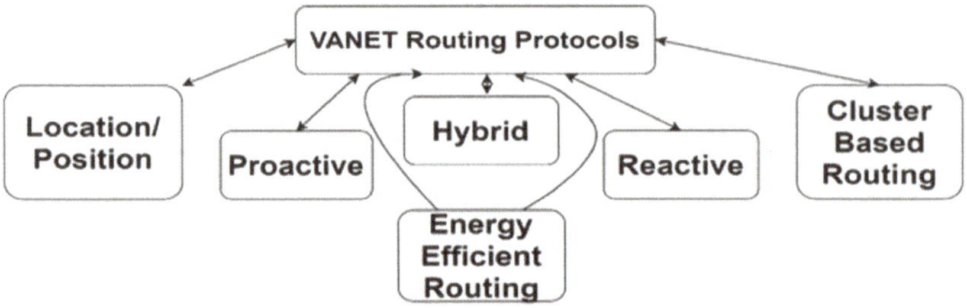

Figure 7: VANET-based routing protocols.

7 ML/DL techniques

Machine learning techniques can be used in VANETs. For monitoring, pollution control, and crowdsourcing, big datasets are utilized to improve the entire network. Here, machine learning is divided into three main subclasses [25].

Supervised learning
Using preexisting dataset to train algorithms in VANETs [26], the commonly used techniques like k-nearest neighbors [27], support vector machine [28], and decision tree [29] can be used to improve QoS in VANETs.

Unsupervised learning
These algorithms are only used to deal with unlabeled data. Unsupervised learning can be used especially in clustering techniques [26].

Reinforcement learning

Reinforcement is usually applied where agent uses to learn directly from the environment. These techniques can be used in gaming and route finding [26]. Figure 8 represents the learning mechanism in agent-based reinforcement. Deep reinforcement learning is used to have reward, offloading, and observed state, where RSU plays an important role.

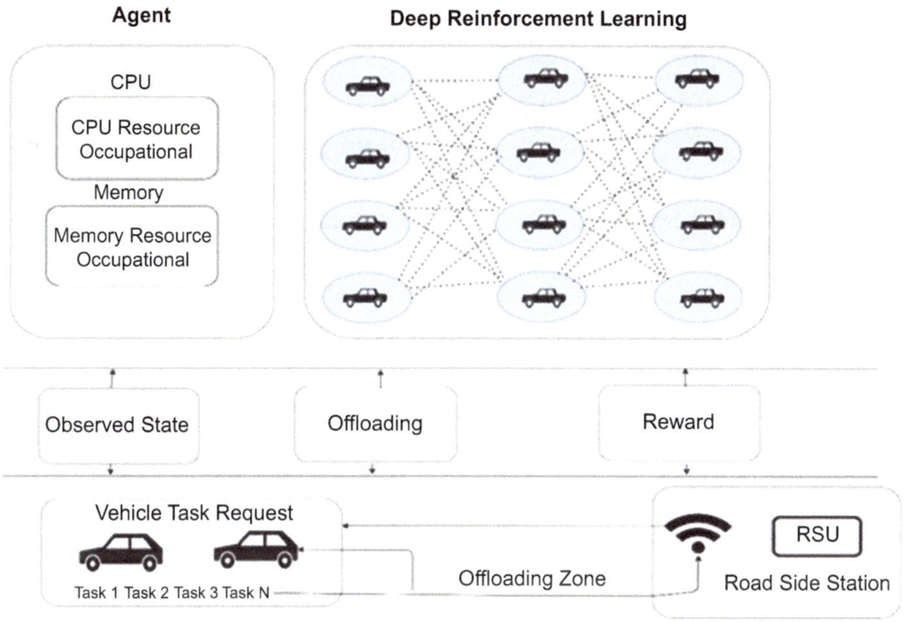

Figure 8: Agent-based deep reinforcement learning.

8 Evolutionary computing algorithms

8.1 Particle swarm optimization routing

Particle swarm optimization routing is proposed in VANETs which use position-based routing. Global positioning system tries to identify the location of each vehicle. This concept is very helpful in smart cities environment. A swarm of vehicles that are used to initialize distinct location must include velocity and direction for each solution. Position and velocity updated regularly in particle swarm optimization. Also, fitness function is calculated. However, if the value of fitness will be greater, then it will be called global optimum [30].

8.2 Genetic routing

Random search genetic routing can be used to improve the natural selection process. Genetic algorithm simulates population for abstract illustration. In the biological part, chromosomes can be used to make candidate solution. This algorithm helps in random searching to optimize artificial agents in sophisticated way [31].

8.3 AntHocNet routing

AntHocNet routing is a natured-inspired algorithm which improves the computational cost. Ant colony optimization–based routing in VANETs can be used in multiobjective approaches. Ants are used to find the best possible route from source to destination. AntHocNet is basically a hybrid technique which has proactive and reactive features [32].

9 Intrusion detection system for VANETs

Intrusion detection system is used in many networks which help detect security attacks. Monitoring of unwanted data packets is used to identify abnormal activities and send information to the base station [33]. There are three types of IDS.

Anomaly-based IDS: this type of IDS mainly focuses on the formulation of novel concept of threshold. Different parameters of QoS need to be considered for the detection of security attacks.

Signature-based IDS: signature-based IDS tried to have a database which stores some signatures of cyber-attacks.

Hybrid IDS: this will have features of both anomaly and signature.

Figure 9 gives the illustration of intrusion detection system in VANETs. The topological scenario consists of cloud, firewall, VANET network, and IDS. Therefore, IDS algorithm will play an important role to identify different security attacks.

10 Conclusion and future challenges

VANETs are commonly called ground-based networks. This survey is giving the concept of VANETs and improving routing protocols, and wireless communication technologies use different techniques like machine learning, deep learning, and heuristic computations. Also, in security point of view, intrusion detection system is studied especially for VANETs to detect security attacks. Comprehensive comparative

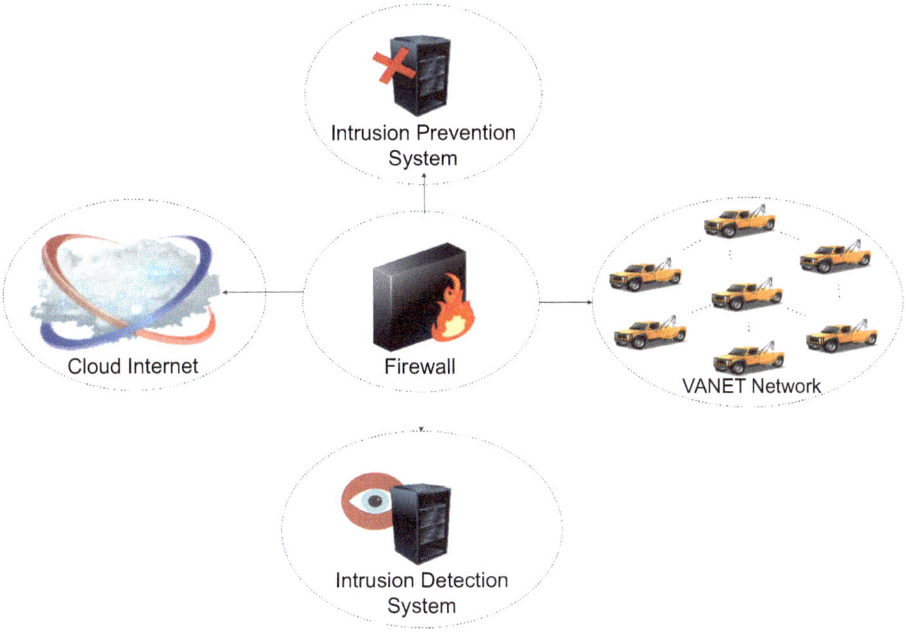

Figure 9: Concept of intrusion detection system in VANETs.

analysis is being performed in wireless networks from 1G to 6G. Ad hoc networks like MANETs, VANETs, and flying ad hoc networks are taken into consideration.

In future, metaheuristic algorithms can be utilized to improve network performance in VANETs. Also, integration of aerial vehicles with VANETs will reduce end-to-end delay. In addition, this survey chapter will help engineers, professionals, researchers, and scientists to work more toward VANETs.

References

[1] Alves Junior J, Wille ECG. Routing in vehicular ad hoc networks: Main characteristics and tendencies. *Journal of Computer Networks and Communications*, 2018:2018.

[2] Yousefi S, Mousavi MS, Fathy M. Vehicular ad hoc networks (VANETs): Challenges and perspectives. *2006 6th International Conference on ITS Telecommunications*. 761–766, 2006. 10.1109/ITST.2006.289012.

[3] Sadakale R, Ramesh NVK, Patil R. TAD-HOC routing protocol for efficient VANET and infrastructure-oriented communication network. *Journal of Engineering*, 2020:2020.

[4] Broustis I, Faloutsos M. Routing in vehicular networks: Feasibility, modeling, and security. *International Journal of Vehicular Technology*, 2008:2008.

[5] Afzal K, Tariq R, Aadil F, Iqbal Z, Ali N, Sajid M. An optimized and efficient routing protocol application for IoV. *Mathematical Problems in Engineering*, 2021:2021.

[6] Ott GD. Vehicle location in cellular mobile radio systems. *IEEE Transactions on Vehicular Technology*, 26:43–46, 1977.
[7] Shah AFMS, et al. Survey and performance evaluation of multiple access schemes for next-generation wireless communication systems. *IEEE Access*, 2021.
[8] Giordani M, et al. Toward 6G networks: Use cases and technologies. *IEEE Communications Magazine*, 58(3):55–61, 2020.
[9] Del Peral-rosado JA, et al. Survey of cellular mobile radio localization methods: From 1G to 5G. *IEEE Communications Surveys & Tutorials*, 20(2):1124–1148, 2017.
[10] Zhao Y. Standardization of mobile phone positioning for 3G systems. *IEEE Communications Magazine*, 40:108–116, 2002.
[11] Spirito MA. On the accuracy of cellular mobile station location estimation. *IEEE Transactions on Vehicular Technology*, 50:674–685, 2001.
[12] Khan IU, Qureshi IM, Aziz MA, Cheema TA, Shah SBH. Smart IoT control-based nature inspired energy efficient routing protocol for flying ad hoc network (FANET). *IEEE Access*, 8: 56371–56378, 2020. 10.1109/ACCESS.2020.2981531.
[13] Khan IU, Alturki R, Alyamani HJ, Ikram MA, Aziz MA, Hoang VT, Cheema TA. RSSI-controlled long-range communication in secured IoT-enabled unmanned aerial vehicles. *Mobile Information Systems*, 2021:2021.
[14] Shah SB, Zhe C, Yin F, Khan IU, Begum S, Faheem M, Khan FA. 3D weighted centroid algorithm & RSSI ranging model strategy for node localization in WSN based on smart devices. *Sustainable Cities and Society*, 39:298–308, 2018.
[15] Khan IU, Hassan MA, Alshehri MD, Ikram MA, Alyamani HJ, Alturki R, Hoang VT. Monitoring system-based flying IoT in public health and sports using ant-enabled energy-aware routing. *Journal of Healthcare Engineering*, 2021:2021.
[16] Hassan MA, Ullah SI, Salam A, Ullah AW, Imad M, Ullah F. Energy efficient hierarchical based fish eye state routing protocol for flying ad-hoc networks. *Indonesian Journal of Electrical Engineering and Computer Science*, 21(1):2021, 465–471.
[17] Al-Absi MA, et al. Moving ad hoc networks – A comparative study. *Sustainability*, 13(11):6187, 2021.
[18] Ismail DPII, Ja'afar MHF. Mobile ad hoc network overview. *2007 Asia-Pacific Conference on Applied Electromagnetics*. 1–8, 2007. 10.1109/APACE.2007.4603864.
[19] Hassan MA, Ullah SI, Khan IU, Hussain Shah SB, Salam A, Ullah Khan AW. Unmanned aerial vehicles routing formation using fisheye state routing for flying ad-hoc networks In *The 4th International Conference on Future Networks and Distributed Systems (ICFNDS)*. 1–7, 2020.
[20] Naseer Qureshi K, Bashir F, Iqbal S. Cloud computing model for vehicular ad hoc networks. *2018 IEEE 7th International Conference on Cloud Networking (CloudNet)*. 1–3, 2018, 10.1109/CloudNet.2018.8549536.
[21] Bitam S, Mellouk A, Zeadally S. VANET-cloud: A generic cloud computing model for vehicular ad hoc networks. *IEEE Wireless Communications*, 22(1):2015, 96–102.
[22] Yadav S, Rajput NK, Sagar AK, Maheshwari D. Secure and reliable routing protocols for VANETs. *2018 4th International Conference on Computing Communication and Automation (ICCCA)*. 1–5, 2018, 10.1109/CCAA.2018.8777690.
[23] Kakarla J, Siva Sathya S, Govinda Laxmi B. A Survey on Routing Protocols and its Issues in VANET. 2011.
[24] Toulni H, Nsiri B. A hybrid routing protocol for VANET using ontology. *Procedia Computer Science*, 73:94–101, 2015.
[25] Khatri S, Vachhani H, Shah S, Bhatia J, Chaturvedi M, Tanwar S, Kumar N. Machine learning models and techniques for VANET based traffic management: Implementation issues and challenges. *Peer-to-Peer Networking and Applications*, 14(3):2021, 1778–1805.

[26] Gillani M, Niaz HA, Tayyab M. Role of machine learning in WSN and VANETs. *International Journal of Electrical and Computer Engineering Research*, 1(1):2021, 15–20.

[27] Beyer K, Goldstein J, Ramakrishnan R, Shaft U. When is "nearest neighbor" meaningful?. In *International conference on database theory*. Springer, Berlin, Heidelberg, 217–235, 1999.

[28] Luong NC, Hoang DT, Gong S, Niyato D, Wang P, Liang Y-C, In Kim D. Applications of deep reinforcement learning in communications and networking: A survey. *IEEE Communications Surveys & Tutorials*, 21(4):2019, 3133–3174.

[29] Amodei D, Olah C, Steinhardt J, Christiano P, Schulman J, Dan M. Concrete problems in AI safety. *arXiv preprint arXiv:1606.06565*. 2016.

[30] Yelure B, Sonavane S. Particle swarm optimization based routing method for vehicular ad-hoc network. *2020 International Conference on Communication and Signal Processing (ICCSP)*. 1573–1578, 2020. 10.1109/ICCSP48568.2020.9182305.

[31] Zhang G, Wu M, Duan W, Huang X. Genetic algorithm based QoS perception routing protocol for VANETs. *Wireless Communications and Mobile Computing*, 2018:2018.

[32] Melaouene N, Romadi R. An enhanced routing algorithm using ant colony optimization and VANET infrastructure. In *MATEC Web of Conferences*. Vol. 259, EDP Sciences, 02009, 2019.

[33] Abdollahi A, Fathi M. An intrusion detection system on ping of death attacks in IoT networks. In *Wireless Personal Communications*. 1–14, 2020.

Meriem Houmer, Mariya Ouaissa, Mariyam Ouaissa,
Siham Eddamiri

Applying machine learning algorithms to improve intrusion detection system in IoV

Abstract: Since the last decade, Internet of vehicles (IoV) keeps attracting the interest of researchers and automakers with a growing variety of applications and services. The aim of these applications is to provide road safety and comfort for drivers and passengers. However, the quantities of data propagated in this network make its security a challenge. The big data generated, collected, and processed on IoV involve valuable and delicate information that malicious intruders might manipulate. Besides that, the specific characteristics of IoV make intrusion detection in this network complicated. The intrusion detection system (IDS) represents the most common approach for network protection. It monitors the traffic activity to identify signs of safety risks and generates alerts for any security anomalies detected.

In order to ensure the accuracy of intrusion detection in IoV, the main purpose of this chapter is to apply machine learning (ML) algorithms for training the IDS with a sufficient dataset of security menaces. The IDS based on ML will be able to make crucial decisions in case of intrusion detection whereas continuing to learn about their highly dynamic environment. This chapter describes different security issues in IoV and presents diverse ML algorithms employed for constructing IDS toward protecting IoV from diverse cyber-attacks.

Keywords: IoV, security, big data, IDS, machine learning

1 Introduction

Wireless networks have revolutionized data exchanges and defined a new paradigm of "always on–always connected." Within this paradigm, communications in vehicular environments open a new field of research in the scientific community. The most common way of looking at wireless networks is one in which mobile clients connect to a base station (BS) that controls communications. This BS covers a certain

Meriem Houmer, High School of Technology - Dakhla, Ibn Zohr University, Morocco,
e-mail: houmer.m@gmail.com
Mariya Ouaissa, Moulay Ismail University Meknes, Morocco, e-mail: mariya.ouaissa@edu.umi.ac.ma
Mariyam Ouaissa, Moulay Ismail University Meknes, Morocco, e-mail: mariyam.ouaissa@edu.umi.ac.ma
Siham Eddamiri, Department of Mathematics and Computer Science, ENSAM, Moulay Ismail University,
Meknes, Morocco, e-mail: sihameddamiri@gmail.com

https://doi.org/10.1515/9783110760477-003

coverage area in which all clients they control can communicate with each other. The outreach to clients of other networks is done through a generally fixed segment. Clients are able to move around and switch BSs without coverage outage through a process called handover.

Ad hoc communications and more precisely vehicular ad hoc network (VANET) or Internet of vehicle (IoV) communications pose new challenges. In this type of network, there is no network infrastructure but rather it is made up of the autonomous mobile nodes themselves communicating with each other over wireless links.

This environment eliminates the centralized control of the network provided by the BS. The nodes must assume responsibilities for routing and maintaining the network. Network control is distributed among the nodes themselves.

As distributed systems, the VANET faces many challenges in terms of security and confidentiality, due to the fact that the broadcast data is broadcasted in an open-access environment. Specifically, to achieve the primary objective of applying IoV, especially those of road safety applications, security provisioning is necessary to ensure the proper functioning of IoV. Otherwise, IoV could become a practical tool to undermine safety and traffic management. For example, by broadcasting false warnings or alert messages with falsified content, the attacker can cause a collision between vehicles. It can even prevent packet routing by attacking network availability in order to make that network or one of its services unavailable, and then its main purpose will become useless.

Several techniques have been proposed to overcome IoV security vulnerabilities. Intrusion detection system (IDS) is an active process used in IoV to analyze network activity and examine traffic data to distinguish unauthorized entry or malicious activity. Currently, machine learning (ML) technology plays an important role in the IoV security by developing security detection models. Different ML algorithms are adapted and involved in IDS in order to analyze traffic data, identify the type of attacks, and then detect the malicious nodes from benign ones. This chapter describes different security issues in IoV and presents various ML algorithms utilized for constructing IDS toward protecting IoV from diverse cyber-attacks.

The rest of this chapter is structured as follows: Section 2 presents the background includes information about vehicles and the IoV architecture and applications. Section 3 describes the security issues in IoV. Section 4 highlights the access control mechanism to access the network and its services. Section 5 illustrates the IDS architecture and techniques. Section 6 explores different ML algorithms involved in IDS in IoV. In Section 8, we conclude the chapter.

2 Background

2.1 Human–machine interface for vehicle

Currently, comfort settings within a vehicle are made through one or more controls. Each of the control has its specific function, for example, for air conditioning temperature and mode, seat heater intensity, audio volume, audio source, and glasses. In addition to having the comfort configuration controls, the vehicles have controls for other vehicle configurations, either security or control; for example, light control, driving assistants such as cruise control, mirror position, and locks.

The types of controls that exist in today's vehicles are:

- The on/off buttons are probably the most common and simple of the controls, but at the same time, their functionality is limited.
- The rotary knobs can be multifunctional, making them a bit more complex and requiring more user attention to configure.
- Touchscreens are becoming increasingly popular in vehicles as they provide versatility; however, that adds complexity.
- Gesture control is mostly associated with touch screens.
- Voice control is less popular than the previous ones.

A driver receives information from the environment, the vehicle, and social media; so effective display of driving and entertainment information is necessary to focus his/her attention on the road. At the same time, it must have an optimized human–machine interface (HMI) to map all the functionalities of a vehicle with accessibility [1].

Despite this, some of the biggest changes regarding the HMI in vehicles were the steering wheel controls, touch screens, center console, and voice control. Furthermore, the biggest challenges for the next few years are the integration of multiple assistance systems, driver distractions, driver information overload, safety verification, and improved comfort. Therefore, some of the research areas with the greatest potential are the needs of the driver, improvements in the technology used in HMI, and management of the resources of HMI [2].

Technological advances have enabled the development of driver–vehicle interface (DVI) systems with the purpose of improving the driving experience in vehicles [3]. The goal of DVI is to improve passenger comfort and limit operator distraction. Divergence of driver attention increases the risk of collision. Distraction is caused when the driver pays attention to other activities, such as the multimedia systems in the vehicle or the use of mobile devices. Feedback between driver and vehicle and design alterations showed a positive influence on driver behavior. Visual feedback from the oncoming traffic has led to improved vehicle performance [4]. Current research in augmented reality applications will possibly provide methods for the future of DVI.

Current research suggests that the mode of transportation has a significant effect on passenger feelings. Additionally, in-vehicle entertainment, such as music, has been shown to improve driver performance and comfort. The significance of the role that DVI systems play in vehicles will increase as human control activities are minimized. Rather than preventing distractions, the current DVI is predicted to evolve to be used to increase passenger productivity and provide feedback on current maneuvers to improve passenger comfort.

2.2 Information generated in a vehicle

The use of electronic systems in vehicle design and development has increased in recent years. The current trend indicates that it will continue to increase since electronic control systems are cheaper, lighter, and faster to implement compared to a mechanical or hydraulic system that provides the same functionality. As well, the number of electronic control units (ECU) inside a vehicle will continue to increase.

An ECU inside a vehicle generates information corresponding to the functionality it implements and this information is shared with other ECU. A vehicle contains one or more networks of interconnected ECU, and through these networks the information generated by each one is exchanged. The information is only shared with the ECU that needs it and this dependency is defined from the network design.

The information transmitted by an ECU can be the result of internal data processing or come from an external source, such as a sensor. The following list names some examples of the information generated in a vehicle:
- Temperature inside/outside the vehicle
- AC/heating mode (face, feet, defroster)
- AC temperature
- Heating temperature
- Seat heating temperature/level
- Emergency call
- Car speed
- Battery level
- Fuel level
- Cruise control (speed, enabled)
- Tire pressure
- Cluster configuration
- GPS information
- Multimedia system

2.3 Internet of vehicles

Currently, the number of vehicles has grown due to the growth in urban population and the rapid expansion of cities [5]. Electric cars (EVs), both completely electric and hybrid, have also been increased. Considering their high mobility, better communication and interconnectivity between these vehicles are needed.

As vehicles have evolved from simple means of transportation to intelligent entities with census and communication capabilities, they have been an essential element of smart cities.

Smart vehicles have five characteristics: autonomous, safe driving, social driving, electric vehicles, and mobile applications.

The Internet of things (IoT) is a vast network connecting intelligent objects and allowing them to interact with each other. When these intelligent devices linked to the Internet represent vehicles, the IoT is becoming the IoV. Therefore, IoV is a widespread application of IoT in the area of smart transportation. Its aim is to act as a crucial data sensor and processing platform for intelligent transport systems (ITS). The vehicle is a census platform, which absorbs environmental information from other vehicles and drivers, and employs it to ensure safe driving, pollution control, and traffic monitoring.

IoV consists of vehicles communicating with each other, as well as with pedestrian devices and units on the roads through vehicle to vehicle (V2V), vehicle to road, vehicle to human, and vehicle to sensor (V2S) to create a social network where the participants are smart devices instead of human beings. All these communications can be grouped in vehicle-to-X communication [6], where X is vehicles, road side units, humans, and the Internet. This refers to the convergence of social IoV (SIoV). SIoV is basically a vehicular instance of the IoT. IoV can be considered as a set of VANET originating from the mobile ad hoc network (MANET). IoV extends the scale, structure, and application of VANET.

Vehicles in IoV use radio frequency identification (RFID) and sensors to collect a large amount of data and analyze it. The VANETs are the source of the vehicle communication in IoV. Moreover, onboard units and smartphones have been added and work collaboratively with VANET's infrastructure to monitor traffic, share vehicle information, and EV charging.

2.3.1 IoV architecture

There are three common IoV architectures [5] (as shown in Figure 1):

1. This layer includes all the sensors in a vehicle for collection of environmental data and detection of special events such as driving patterns and environmental conditions. In addition, it has RFID, satellite positioning, road environment, vehicle position, perception of other cars and objects, and so on.

2. Network layer is the communication layer that guarantees access to networks through GSM, 5G, WiMax, WLAN, Wi-Fi, and Bluetooth. It can support several wireless communication modes including V2V, vehicle to infrastructure, vehicle to pedestrian, and V2S.
3. Application layer contains statistical tools and supports storage and processing infrastructure. It stores, analyzes, processes, and makes decision on risky situations such as traffic congestions and poor weather. Smart applications, traffic safety, efficiency, and multimedia based on infotainment are all represented in this layer.

For vehicle representation on Internet, each vehicle must have a unique Internet identifier. A global terminal ID (GID) is the core of IoV. The GID provides vehicles with a cyber-tag or cyber-ID.

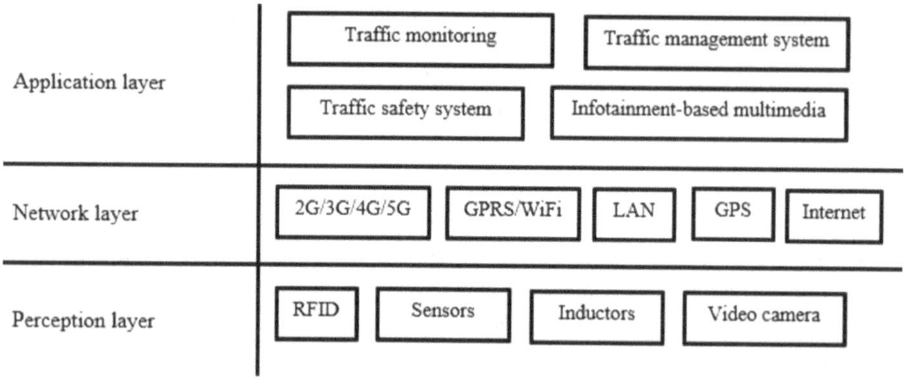

Figure 1: IoV architecture [5].

2.3.2 IoV applications

IoV has a variety of applications, including:

1. Safe driving: it represents cooperative accident prevention systems that utilize sensors to identify and warn the driver of potential accidents. Periodic status and emergency messages are included in this application. An emergency message is triggered by an emergency event like traffic congestion, accident, or poor road condition.
2. Traffic control: IoV provides significant changes for the management of urban congestion, transportation and logistics, vehicular traffic, and a collective lifestyle.
3. Accident response: connected vehicles can automatically send real-time accident information and their location to emergency teams. It could save lives by speeding up response time to an emergency.

4. Convenient services: services such as remotely unlocking a door or retrieving a stolen vehicle have become possible because of the capacity to remotely access a vehicle. Transportation authorities can use connected vehicle technologies to get real-time traffic information and parking information, allowing the management of transportation systems to be more efficient and reduce traffic and vehicular congestion.
5. Infotainment: connected vehicles are capable of providing online options for in-dash entertainment.

Other applications can be electronic booths, traffic guidance systems for safe navigation, autonomous vehicles, intelligent vehicle control, accident prevention, traffic flow monitoring, and autonomy of a vehicle.

3 Security issues in IoV

Like any deployment of the most advanced technological activity, the context of connected vehicles/IoV also presents significant challenges around vulnerabilities and threats. Certainly, IoV are prone to many attack scenarios [7, 8] due to the decentralized nature of the network and the lack of authentication of nodes participating in communications.

3.1 Attackers classification

In order to be able to secure a system, it is essential to identify the nature of its attackers. In ad hoc vehicle networks, attackers can be classified according to the following measures:

- The internal or external attacker: the internal attacker is considered a regular member of the network. It is able to communicate with the other nodes of the network and also to access its various services. The presence of insider attacks is extremely dangerous and difficult to be detected, since it negates the level of protection offered by cryptographic methods. The external attacker is treated as an intruder by the member nodes, and then he/she is limited in the types of attacks he/she can provoke [9].
- The malicious or rational attacker: a malicious attacker has no personal interest through his/her attacks; his/her main objective is to make the network unavailable. Yet he/she can use any means without considering the costs and consequences. On the other hand, a rational attacker pursues his/her personal interests, and therefore one can anticipate the objectives of the attack and the means used [10].

- The passive or active attacker: the passive attacker tunes in illegally, simply monitoring the dissemination or collection of information. The passive attack is difficult to be detected because it does not require any changes in data or system resources. On the other hand, the active attacker tries to modify the information which is disseminated in the network. He/she can modify it, falsify it, or even destroy it [11].

3.2 Attacks in IoV

In IoV security, it is important to define the requirements that the system must meet in order for the network to function properly. Authentication, confidentiality, integrity, and availability of information are the main security requirements to ensure in order to cope with security threats. The failure to meet these requirements can lead to potential attacks or threats in the network.

In IoV, a network node can act maliciously in different ways, and each of which can be classified as an information manipulation attempt.

For example, each node within the network can serve as a router to transmit information between two nodes; however, when a node decides, for some unknown reason, not to retransmit the data packets requested, the information will become unavailable and the availability of the network and its services will be affected. In this case, the safety of users, drivers, and pedestrians is at risk. On the other hand, the confidentiality of the data packets may be compromised when there is an insecure wireless channel with inefficient encryption [12]. Similarly, the integrity of the information is compromised when an external attacker can alter, modify, and even delete data packets transmitted between various elements of the network. Moreover, ensuring authentication in IoV consists in protecting the authentic nodes from outside or inside attackers by filtering the network of nodes using forged identities.

There are different threat vectors that can affect the IoV, such as denial of service attacks, viruses, black hole, and man-in-the-middle attacks [15].

Denial-of-service attacks affect the availability of networks since one or several attackers send packets to saturate a server and leave the network without service. Dasari [16] proposed a detection of attacks at the MAC layer (medium access control) for wireless networks, and thus avoid denial of service attacks. In [17] they discuss how critical security is in a wireless sensor network and highlight several techniques to detect and prevent black hole attacks, proposing a model to conserve energy, in [18] a model to prevent attacks is also proposed of black holes in wireless sensor networks, but they base the model on the knowledge-based learning technique to detect and mitigate harmful nodes. Man-in-the-middle attacks affect the confidentiality of the network since sent messages can be read and spied on by intruders [19]. Message manipulation attacks affect the integrity of the network; an attacker can manipulate the message and deliver an altered and therefore incorrect message to the recipient [20]. Figure 2 represents a summary of the most common attacks on IoV.

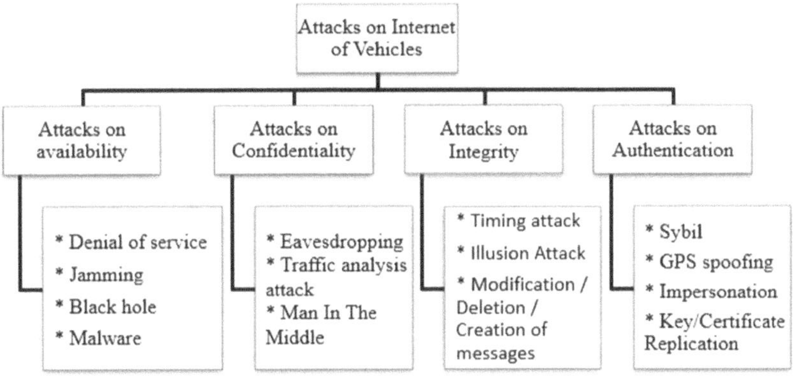

Figure 2: Attacks on IoV.

4 Access control

As in traditional networks, IoV needs a mechanism that controls access to the network and the services it provides. The consequences of an attack in which an intruder would have access to network services can be catastrophic since in IoV the nodes assume management and routing tasks as they do not have a centralization unit. An intruder could divert traffic during routing or gain access to identification keys.

At the network layer, it is necessary to ensure that no unauthorized node joins the network either to receive information or to route it. Likewise, at the application level, it is also essential to ensure that unauthorized elements do not access services, for example, the key management service.

Access control generally consists of authenticating network users. In other words, to access the network and its services, a user must uniquely identify himself and the network authenticates him as authorized for access. In certain ad hoc networks, the services are centralized while in others they are distributed, this fact makes it necessary to use different access control mechanisms. If we choose a distributed access control mechanism for the network, access control based on digital certificates and certificate authorities will be necessary. In other schemes with centralized services, authentication based on user and password is required. It is very useful to carry out a preliminary study of the security needs of the network to be deployed, in this way, the network access control mechanisms can be adapted.

5 Intrusion detection system

Access control consists of the first line of defense to prevent intruders from accessing the network. Indeed, IDS are often employed as a second line of defense.

An IDS is a mechanism that analyzes network activity in order to detect abnormal or suspicious activity. In addition, it tries to delete them by suggesting corrective actions or by triggering actions on events (alerts, mail, reset the connection . . .). In IoV, IDSs are responsible for analyzing incoming and outgoing packets to detect malicious signatures. Basically, the IDSs can be classified according to the detection techniques used in the system or according to its architecture.

5.1 Intrusion detection techniques

- Signature-based system: It analyzes user behavior and compares it to known intrusion patterns (called signatures) [21]. The system consists of an internal database of signatures. An alarm will be raised when the activity of a user fits recorded patterns/signatures.
- Anomaly detection system: It compares system activities with that of regular behavior and raises an alarm when a deviation from normal conduct exceeds a preset threshold. It is divided into two parts: training and detection [22]. Before it can be utilized in a detection model, it must be trained on typical behavior. Using heuristic approaches or rules distinguishes aberrant from typical activity in detection mode.
- Specification-based system: The system defines a set of constraints that describe the proper functioning of a protocol or program and supervises the protocol at a time with the constraints defined to identify deviations [23]. Malicious activity is detected if the program or protocol does not meet the established conditions for proper operation.
- Hybrid or compound IDS: It is a combination of two or more detection techniques [24].

5.2 Intrusion detection architectures

- Autonomous IDS: In this architecture, an autonomous intrusion detection engine is installed on each node to perform intrusion detection independently and make the decision on malicious activities based on the data collected by the node [25]. Therefore, no data is exchanged. In addition, each node has no information on the position of other nodes and no alert information crosses the network.
- Cooperative and Distributed IDS [26]: Cooperative IDS are characterized by cooperation between neighboring nodes. In this architecture, an intrusion detection

engine is installed on each node, which monitors local audit data and provides intrusion detection. They also participate in the cooperative intrusion detection and response mechanism by exchanging audit data and detection results with neighboring nodes.
- Hierarchical IDS: This is a multi-layered architecture that divides the network into groups (clusters). Specific nodes, determined by a cooperative algorithm between nodes, are selected to act as cluster-heads. In this architecture, cooperation is achieved between the group leader and each of the members of the same group, as is the case in ad-hoc multilayer networks.

There are several IDS proposals for vehicular network such:

In [13] a distributed system based on mobile agent technology is proposed. A mobile agent is defined as an autonomous, lightweight, and dynamically updateable software entity that traverses the network and runs on certain nodes. This method is especially appropriate in the case of VANETs, where resources such as the bandwidth of the links or the capacity of the nodes can be limited. The different functions of the IDS are distributed among different types of agents so that the load introduced by the IDS is distributed efficiently between the nodes of the network.

In [14] the authors propose a distributed and cooperative architecture for intrusion detection. In this system, each node runs an IDS agent that monitors the activities locally. If the IDS detect an intrusion from the local traces, it initiates a response procedure. If an anomaly is detected but there is no formal evidence of the intrusion, a cooperative protocol is used with the neighbors to determine if the intrusion took place or not.

Recently, advanced ML algorithms are implied in the IDS in order to analyze traffic data and distinguish the type of attacks. These IDS-based learning algorithms employ the intelligence to construct a model that classifies and differentiates between benign and malign nodes.

6 Machine learning for intrusion detection system in IoV

Nowadays, ML becomes the trendiest science topic in several fields. This technology can be described as the field of study that allows computers to learn without being explicitly programmed. Actually, there is a great diversity of ML algorithms and methods. These are used in a wide variety of applications such as e-commerce, entertainment, medical, banking, research, IoV, and security.

The IDS is a security application in which ML methods are widely used in its design. The challenge of these IDS is to collect and classify diverse behaviors in IoV in an effective manner. To respond to this challenge, several supervised (classification) and unsupervised learning (clustering) algorithms for detecting malicious activities are used. In the following, a variety of ML algorithms applied to the IDS are discussed. These algorithms (supervised and unsupervised algorithms) allow designing IDS with robust detection and low false reports. Moreover, they allow the system that adjusts fast to the changing malicious behavior.

6.1 The supervised algorithms

- The ID3 algorithm: ID3 is a supervised classification algorithm. It relies on examples already classified in a set of classes to establish a classification model. The model that ID3 produces is a decision tree. This tree will be used to classify new samples.

  ID3 recursively builds a decision tree by choosing the attribute that maximizes information gain according to Shannon's entropy. At each step of the recursion, it calculates from among the attributes remaining for the current branch, the one that will maximize the information gain. This is the attribute that will most easily classify examples at this level of this branch of the tree.

  Another prerequisite for the ID3 algorithm is that the fields are all discrete. It is therefore interesting to discretize the continuous fields according to the values of the < target > field before launching the algorithm. Often the target attribute only takes the values true, false or fail, and success.
- k-Nearest neighbor (KNN): KNN is a standard classification algorithm which is based exclusively on the choice of the classification metric. It is nonparametric (only k needs to be set) and is based only on training data.

  In KNN, from a labeled database, we can estimate the class of a new data item by identifying the majority class of the k closest neighboring data.

  To provide a prediction, the KNN method is based on the whole dataset. In fact, if an observation is not part of the dataset which we wish to forecast, the algorithm will search for K instances of the dataset nearest to our observation. For these K neighbors, the algorithm will then calculate the values of the variable y for the observation we want to forecast, based on their output variables y.

  Furthermore:
  - When KNN is utilized for regression, the mean (or median) of the y variables of the K closest observations must be employed for the prediction.
  - When KNN is utilized for classification, the mode of the y variables of the K closest observations must be employed for the prediction.
- Multilayer perceptron (MLP): An MLP is a neural network (inspired by the visual system) connecting several layers in a directed graph, which means that the signal path through the nodes is only one way.

It has a layer dedicated to perception and a layer dedicated to decision making. The specificity of a neural network is therefore not in its computers, which represent only a series of equations. It is found in the links between the calculators.

In MPL, each node, except input nodes, has a nonlinear activation function. An MLP uses back-propagation as a supervised learning technique. Since there are multiple layers of neurons, MLP is a deep learning technique.

- Support vector machines (SVM): SVM is an algorithm that solves classification, regression, or anomaly detection problems. The purpose of SVM is to separate data into classes using a boundary so that the distance between different groups of data and the boundary between them is maximum.

  In order for the SVM to find this boundary, it is necessary to give the training data. The SVM will predict the most likely position of the boundary based on this data; this is the training phase required for any algorithm for ML.

  Once the training phase was completed, the SVM found, from training data, the supposed location of the border. In this way, he/she learned the location of the border from the training data.

6.2 The unsupervised algorithms

- K-means: These are unsupervised ML algorithms that are employed to overcome clustering problems. They divide and classify a set of unlabeled data points (without external classification) into a cluster (unrelated to server clusters). Each algorithm iteration assigns a set of similar characteristics to each point. Data points may be traced through time to detect changes occurring in clusters. K-means algorithms can validate hypotheses about the types of groups that exist in a specific dataset, or be employed to discover unknown clusters.
- Single-linkage clustering: In this method, the distance between two clusters is represented by the minimum distance between all pairs of data between these two clusters (pair composed of one element of each cluster), we are talking about the minimum hop. However, we choose as the representative of the group the element which is the most similar to the elements of the other group. Thus, the similarity between clusters is the maximum of the similarity between the elements of the clusters.

  The strong point of this approach is that it is very good at detecting elongated classes, but its weak point is that it is sensitive to the chain effect, and therefore less suitable for detecting spherical classes.
- Quarter-sphere SVM (QS-SVM) [27] were created to deal with intrusion detection data that have one-sided features centered around the origin. QS-SVM is based on the notion of adopting a sphere over the data mass center. The distance of a data point from the center of the sphere defines an anomaly score. The radius of the sphere with normal data points is defined by a threshold for the attack scores.

According to the literature, different ML algorithms have been used in IDS in vehicular environments to detect diverse attacks and filter out malicious nodes by analyzing data collected from networks. Table 1 presents some IDS-based ML algorithms. However, several simulations of the proposed IDS-based ML, in the literature, are often carried out using small data generated by simulators and just few methods utilize intrusion detection databases containing real attack scenarios. Still, it is anticipated that the ratio of vehicular data generated will be increasing exponentially.

Table 1: Machine learning for intrusion detection systems in VANET.

Papers	Algorithms used	Description	Intrusion datasets/ simulator
[28]	Recurrent neural network, convolutional neural network	Propose an end-to-end intrusion detection technique to detect malware traffic for on-board units	ISCX2012
[29]	k-Means based on self-organized map	Meet the high accuracy and precision requirements of intrusion detection systems	NSL-KDD
[30]	SVM, neural network	Propose an intelligent IDS to handle gray holes and rushed attacks in self-driving vehicular networks	SUMO
[31]	SVM	Provide a collaborative IDS for the dynamic detection of different sorts of attacks.	KDD Cup 1999
[32]	Logistic regression	Propose a distributed framework for the conception of VANET collaborative detection systems	NSL-KDD
[33]	Decision tree, multilayer perceptron	Propose IDS for DoS and fuzzy attacks	Car hacking dataset
[34]	K-means, linear splines	Propose an IDS model to deal with the dynamic topology of VANETs	NSL-KDD
[35]	Neural networks	Propose a hybrid intrusion detection for connected self-driving vehicles	Kyoto dataset

7 Conclusion

This chapter provides an overview of IoV security and IDS based on ML algorithms in vehicular environments. We give the details of the HMI embedded in vehicles. We present generally the IoV, its architecture, and its application. Then, we have interested in the different attacks that IoV confront; also we introduce the IDS which represents an effective countermeasure against these attacks. Finally, we have focused on ML algorithms applied to the IDS by describing the different ML methods (supervised and unsupervised methods) and presenting some proposed IDS-based ML algorithms in order to enhance the detection of anomalies.

References

[1] Herrmann A, Brenner W, Stadler R. Human–machine interaction. In *Autonomous Driving*. Emerald Publishing Limited, 2018.

[2] Weyer J, Fink RD, Adelt F. Human–machine cooperation in smart cars. An empirical investigation of the loss-of-control thesis. *Safety Science*, 72:199–208, 2015.

[3] Elbanhawi M, Simic M, Jazar R. In the passenger seat: Investigating ride comfort measures in autonomous cars. *IEEE Intelligent Transportation Systems Magazine*, 7(3):4–17, 2015.

[4] Nobari KD, Albers F, Bartsch K, Bertram T. Online feedback control for driver-vehicle interaction in automated driving. In *International Conference on Applied Human Factors and Ergonomics*. Springer, Cham, 159–165, 2020

[5] Sadiku MN, Tembely M, Musa SM. Internet of vehicles: An introduction. *International Journal of Advanced Research in Computer Science and Software Engineering*, 8(1):11, 2018.

[6] Chowdhary N, Kaur PD. Addressing the characteristics of mobility models in IoV for smart city. In *2016 International Conference on Computing, Communication and Automation (ICCCA)*. IEEE, 1298–1303, 2016.

[7] Raya M, Hubaux JP. Securing vehicular ad hoc networks. *Journal of Computer Security*, 15(1): 39–68, 2007.

[8] Hasrouny H, Samhat AE, Bassil C, Laouiti A. VANet security challenges and solutions: A survey. *Vehicular Communications*, 7:7–20, 2017.

[9] Abdel-Fattah F, Farhan KA, Al-Tarawneh FH, AlTamimi F. Security challenges and attacks in dynamic mobile ad hoc networks MANETs. In *2019 IEEE Jordan International Joint Conference on Electrical Engineering and Information Technology (JEEIT)*. IEEE, 28–33, 2019

[10] Kohli P, Painuly S, Matta P, Sharma S . Future trends of security and privacy in next generation VANET. In *2020 3rd International Conference on Intelligent Sustainable Systems (ICISS)*. IEEE, 1372–1375, 2020.

[11] Upadhyaya AN, Shah JS. Attacks on VANET security. *International Journal of Computer Engineering and Technology*, 9(1):8–19, 2018.

[12] Alheeti KMA, Gruebler A, McDonald-Maier KD. An intrusion detection system against malicious attacks on the communication network of driverless cars. In *2015 12th Annual IEEE Consumer Communications and Networking Conference (CCNC)*. IEEE, 916–921, 2015.

[13] Kachirski O, Guha R. Intrusion detection using mobile agents in wireless ad hoc networks. In Proceedings. IEEE Workshop on Knowledge Media Networking. IEEE, 153–158, 2002.

[14] Houmer M, Hasnaoui ML. A hybrid intrusion detection system against egoistic and malicious nodes in VANET. In *The Proceedings of the Third International Conference on Smart City Applications*. Springer, Cham, 961–973, 2019.

[15] Sinha P, Jha VK, Rai AK, Bhushan B. Security vulnerabilities, attacks and countermeasures in wireless sensor networks at various layers of OSI reference model: A survey. In *2017 International Conference on Signal Processing and Communication (ICSPC)*. 288–293, 2017.

[16] Dasari M. Real time detection of MAC layer DoS attacks in IEEE 802.11 wireless networks. In *2017 14th IEEE Annual Consumer Communications Networking Conference (CCNC)*. 939–944, 2017.

[17] Farooq MU, Wang X, Yasrab R, Qaisar S. Energy preserving detection model for collaborative black hole attacks in wireless sensor networks. In *2016 12th International Conference on Mobile Ad-Hoc and Sensor Networks (MSN)*. 395–399, 2016.
[18] Kaur H, Singh A. Identification and mitigation of black hole attack in wireless sensor networks. In *2016 International Conference on Micro-Electronics and Telecommunication Engineering (ICMETE)*. 616–619, 2016.
[19] Bhushan B, Sahoo G, Rai AK. Man-in-the-middle attack in wireless and computer networking – A review. In *2017 3rd International Conference on Advances in Computing, Communication & Automation (ICACCA)(Fall)*. IEEE, 1–6, 2017.
[20] Gupta A, Jha RK. Security threats of wireless networks: A survey. In *International Conference on Computing, Communication Automation*. 389–395, 2015.
[21] Hoang XD, Hu J, Bertok P. A program-based anomaly intrusion detection scheme using multiple detection engines and fuzzy inference. *Journal of Network and Computer Applications*, 32(6):1219–1228, 2009.
[22] Chandola V, Banerjee A, Kumar V. Anomaly detection: A survey. *ACM Computing Surveys (CSUR)*, 41(3):1–58, 2009.
[23] Scarfone K, Mell P. *Guide to Intrusion Detection and Prevention Systems (IDPS)*. National Institute of Standards and Technology, 2012.
[24] Awodele O, Idowu S, Anjorin O, Joshua VJ, Multi-Layered A. Approach to the design of intelligent intrusion detection and prevention system (IIDPS). *Issues in Informing Science & Information Technology*, 6:2009.
[25] Chadli S, Emharraf M, Saber M, et al. Combination of hierarchical and cooperative models of an IDS for MANETs. In *2014 Tenth International Conference on Signal-Image Technology and Internet-Based Systems*. IEEE, 230–236, 2014.
[26] Zhang Y, Lee W. Intrusion detection in wireless ad-hoc networks. In *Proceedings of the 6th annual international conference on Mobile computing and networking*. 275–283, 2000.
[27] Laskov P, Schäfer C, Kotenko I, Müller KR. Intrusion detection in unlabeled data with quarter-sphere support vector machines, 2004
[28] Zeng Y, Qiu M, Zhu D, Xue Z, Xiong J, Liu M . DeepVCM: A deep learning based intrusion detection method in VANET. In *2019 IEEE 5th Intl Conference on Big Data Security on Cloud (BigDataSecurity), IEEE Intl Conference on High Performance and Smart Computing, (HPSC) and IEEE Intl Conference on Intelligent Data and Security (IDS)*. IEEE, 288–293, 2019.
[29] Almi'ani M, Ghazleh AA, Al-Rahayfeh A, Razaque A. Intelligent intrusion detection system using clustered self organized map. In *2018 Fifth international conference on software defined systems (SDS)*. IEEE, 138–144, 2018.
[30] Ali Alheeti KM, Gruebler A, McDonald-Maier K. Intelligent intrusion detection of grey hole and rushing attacks in self-driving vehicular networks. *Computers*, 5(3):16, 2016.
[31] Kim M, Jang I, Choo S, Koo J, Pack S. Collaborative security attack detection in software-defined vehicular networks. In *2017 19th Asia-Pacific Network Operations and Management Symposium (APNOMS)*. IEEE, 19–24, 2017.
[32] Zhang T, Zhu Q. Distributed privacy-preserving collaborative intrusion detection systems for VANETs. *IEEE Transactions on Signal and Information Processing over Networks*, 4(1):148–161, 2018.
[33] Nazakat I, Khurshid K. Intrusion detection system for in-vehicular communication. In *2019 15th International Conference on Emerging Technologies (ICET)*. IEEE, 1–6, 2019.
[34] Schmidt DA, Khan MS, Bennett BT. Spline-based intrusion detection for VANET utilizing knot flow classification. *Internet Technology Letters*, 3(3):e155, 2020.
[35] Alheeti KMA, McDonald-Maier K. Hybrid intrusion detection in connected self-driving vehicles. In *2016 22nd International Conference on Automation and Computing (ICAC)* (pp. 456–461). IEEE, 2016.

T. Chandrakumar, T. Ramya, J. Suganthi

Auto steering control in a self-driving car using deep learning model

Abstract: Man-made brainpower has been seeing a fabulous improvement in defeating any issues between the capacities of individuals and machines. The levels of progress in computer vision with profound learning have been made and satisfied with time, fundamentally more than one unequivocal assessment – a convolutional neural network (CNN). Our methodology is a start to finish framework that is outfitted by CNN to design crude pixels from a solitary forward-looking camera clearly to coordinate requests for guiding. With least arrangement data from individuals, the system sorts out some way to drive in busy time gridlock on neighboring roads with or without way markings and on roadways, that is, we never train models to recognize a way or road. An end-to-end learning prompts better execution and more unassuming structures. Better execution results in light of the fact that the inward fragments self-smooth out to help the overall structure execution, instead of working on human-picked center principles, for example, path area. Such guidelines are normally picked for straightforwardness of human interpretation which does not therefore guarantee the most prominent structure execution. More unassuming associations are possible considering the way that the structure sorts out some way to handle the issue with the inconsequential number of taking care of steps.

Keywords: convolutional neural network (CNN), self-driving car, autonomous vehicles, smart cars, deep learning

1 Introduction

Autonomous or self-driving vehicles have been a subject of expanding interest lately, as numerous organizations are effectively creating related equipment and programming innovations toward completely autonomous driving ability with no human mediation [1]. An autonomous vehicle is equipped with a facility for recognizing its present situation with no human intercession. A human explorer is not expected to accept accountability for the vehicle in any means. This sort of independent vehicle

T. Chandrakumar, Department of Applied Mathematics and Computational Sciences, Thiagarajar College of Engineering, Madurai, India, e-mail: tckcse@tce.edu
T. Ramya, Independent Researcher, Madurai, India, e-mail: ramyathangavel87@gmail.com
J. Suganthi, Department of Information Science and Engineering, T. John Institute of Technology, Bengaluru, India, e-mail: jsuganthi@tjohngroup.com

https://doi.org/10.1515/9783110760477-004

can go anyplace as an ordinary vehicle driven by a human driver. A totally independent vehicle would act normally, careful, and fit for making its own choices.

Self-driving vehicles rely upon sensors, actuators, complex estimations, deep learning (in general, artificial intelligence (AI) systems), and mind-blowing microprocessors to execute the programming. They make and keep a collaborator of their natural components subject to an assortment of sensors planned in various pieces of the vehicle [2]. For example, close by conditions of different vehicles are screened by radar sensors, and traffic signals and street signs are followed and recognized by camcorders which additionally mission the people by their strolling designs [3]. Lidar (light ID and running) sensors skip beats of light off the vehicle's normal parts to really look at distances, perceive street edges, and see way markings [3]. Ultrasonic sensors coordinated with the wheels of the vehicles can perceive the controls while different vehicles leave around the vehicle [4].

Refined programming through cutting-edge AI structures, for example, profound learning or AI modules, by then gauges this considerable data, plots a way, and sends rules to the vehicle's actuators, which control the speed increase, toning down, and organizing. Hard-coded rules, catch revolution computations, perceptive illustration, and article affirmation help the item by sticking to traffic runs and investigating checks.

The gain of using autonomous cars as the situations for comfort and personal satisfaction enhancements are boundless. Primarily, the elders and physically challenged people can enjoy independent journeys anywhere on the Earth [4]. On the off chance that we were at day camp and failed to remember our things, the vehicle could present us the missing things. We could even send our pets to a veterinary arrangement. However, the certified assurance of independent vehicles is the potential for definitely cutting down CO_2 radiations [5]. In another report, specialists have recognized three examples that, at whatever point got at the same time, would deliver the greatest limit of autonomous vehicles: vehicle computerization, vehicle destroy, and ridesharing. By 2050, these "three surprises in metropolitan transportation" could:
- Diminish jamming in traffic (round about 30% less vehicles)
- Reduce transportation expenses by 40% (as far as vehicles, fuel, and framework)
- Further develop walkability and bearableness
- Let loose parking areas for different utilizations (schools, parks, and public venues)
- Lessen metropolitan CO_2 emanations by 80% around the world

The common challenges for completely self-driving autonomous cars are conditions of the weather, traffic rules, accident liability, Lidar and Radar, state regulation, and AI versus emotional intelligence [6]. What happens when an autonomous vehicle drives in strong precipitation? If there is a layer of snow making the rounds, way dividers evaporate. How could the cameras and sensors track way markings, if the markings are muddled by water, oil, ice, or rubbish? Will these vehicles experience inconvenience in burrows or on ranges? What would it be advisable for them to do in seriously stopped up active time gridlock? Will these vehicles be moved to a particular way? Likewise, the thing might be said about the armada of legacy vehicles truly sharing the roads for the going with future years?

Who is responsible for incidents accomplished by a self-driving vehicle? The maker? The human explorer? The most recent systems propose that an absolutely self-regulating level 5 vehicle would not have a dashboard or a controlling wheel, so a human explorer would not have the decision to acknowledge responsibility for the vehicle in a crisis. Lidar is exorbitant and is at this point endeavoring to observe to be some sort of congruence among the reach and objective. If various free vehicles were to drive on a comparable road, would their Lidar signals intrude with one another? Besides, if distinctive radio frequencies are available, will the repeat go to be adequate to assist huge with scaling assembling of independent vehicles? Human drivers rely upon subtle signs and nonverbal correspondence like outwardly associating with walkers or scrutinizing the looks and nonverbal correspondence of various drivers to make split-subsequent options and anticipate rehearsals. Will self-administering vehicles rehash this affiliation? Will they have a comparative life-saving motivation as human drivers?

To avoid such confessions about the self-driving autonomous cars, this chapter has introduced one such thing on dealing with auto steering alignment with high precision on the way to departing the elders or physically challenged. This model has particularly used the deep learning model that recognizes the instant traffic pattern captured by the front camera placed at the front of the car dashboard.

2 Auto steering control method

For the purpose of experimentation, the data has been collected in real time from DARPA Autonomous Vehicle (DAVE-2) which is available publicly. The collected images are preprocessed and modeled with the hybrid deep learning architecture and evaluated using a high volume of test data. The modules of the hybrid deep learning architecture involves

- Data collection
- Data preprocessing
- Data modeling
- Data validation and testing
- Simulation – sanity check

2.1 Data collection

Data was recorded in and around the cities of Rancho Palos Verdes and San Pedro California. The dataset contains 63,000 images approximately that consumes 3.1 GB of storage. Data has been collected by using a car dashcam with labeled steering angles of each image in a text file which is available as a public repository at github [7].

2.2 Data preprocessing

HSV (hue, saturation, and value) is allegedly better than RGB (red, green, blue) for some computer vision applications because it separates brightness from color. So the image is converted from RGB to HSV to account for the lighting differences in images and the data is stored in a pickle format which allows processing and accessing the data much faster and it can be easily converted into a numpy data type which is converted into tensor data type at the time of model running.

2.3 Data modeling

The preprocessed data has been modeled using a hybrid deep learning architecture that consists of convolutional neural networks (CNNs) with various layers of pooling and activations. CNN influences the spatial data and they appropriately arrange the pictures by relaying them for better performance. It utilizes an impromptu design by taking natural information taken from physiological trials performed on the level of cortex. The vision depends on different cortex levels, and every one perceives its increasingly organized data.

To start with, CNN sees single pixels and afterward it perceives straightforward mathematical structures and more components' structures like articles, faces, human bodies, and creatures. The synopsis of the model is given underneath, which is motivated by DAVE-2 of NVIDIA with a little tweaking and adjustment in the layers.

2.4 CNN – convolutional neural network

CNN, presented by Yann LeCun in 1980, is a calculation in deep learning that takes in a data picture, allots significance to alternate points of view/objects in the image, and has the choice to isolate one from the other. CNNs [1] have transformed acknowledgment of examples [8, 9]. Going before the certain allotment of CNNs, most of the examples of acknowledgments were performed using a hidden period of hand-made component mining tracked by a classifier. The forward hop of CNNs is that elements are procured in this manner from arranging models. The CNN approach is particularly marvelous in picture assertion errands, considering the way that the convolution activity gets the 2D considered pictures. In like manner, by using the convolution pieces to really take a look at an entire picture, to some degree, several limits ought to be learned that stood out from irrefutably the quantity of exercises. Numerous researchers enthusiastically track down a colossal investigation in this area to empower machines to consider to be as individuals do, see it in like manner and even use the data for endless endeavors like image and video accreditation, image analysis and classification, media recreation, recommendation systems, and natural language processing.

While CNNs with learned elements have been working together for more than 20 years [10], their determination has exploded over the latest a few years because of two late developments. To begin with, enormous informational collections, for example, the large-scale visual recognition challenge [11] has become beneficial for preparation and approval. Next, the CNN pattern recognition has been executed on the greatly equal design preparing units (GPUs) which colossally speed up learning and deduction.

The term self-driving is frequently utilized reciprocally with autonomous. Notwithstanding, it is a somewhat unique thing. Autonomous vehicle can drive itself in a couple or even all conditions, but a human voyager ought to reliably be accessible and ready to take control. Self-driving cars are the hot spot area in the automobile industry. Currently, we are in level 4 of cars which was achieved by the brand Tesla. Conventional methods use separate machine learning models for object detection, path detection, segmentation, and so on, and combine their outputs to achieve automation.

The instruction sign will be the steering angle with the corresponding images. In this way, we will actually want to make a shrewd framework which is extremely proficient as far as memory utilization. The dataset contains roughly 63,000 pictures and 3.1 GB of data which was recorded around Rancho Palos Verdes and San Pedro California. The images in the dataset are changed over into HSV shading space to represent the distinctions in lighting conditions and are put away in a sticky situation design for quicker preparation and getting to the data or information. Then, at that point the model is prepared on CNN to acquire the guiding point as yield. The yield is given as the contribution to the mechanical steering.

The proposed methodology describes a CNN that goes deeper than the recognition of patterns. It learns the entire strength of the parallel pipeline expected to co ordinate a vehicle. The justification behind this task was refined more than 10 years sooner in a Defense Advanced Research Project Agency (DARPA) seedling project known as DAVE [12] in which a subscale radio control vehicle went through a junk-filled back entrance way. The DAVE was prepared on critical length of human driving in same, now not vague conditions. The game plan information included video from two cameras got together with left and right controlling solicitations from a human leader.

According to different perspectives, DAVE-2 was charged up by the leading work of Jackel et al. [13] which is collected for the ALVINN (autonomous land vehicle in a neural network) structure. It is everything except the beginning of the arranged neural association without a doubt in coordinate a vehicle on open roads. This work differentiates in that 25 years of advances by applying more data and computational ability to the task. Moreover, this inclusion in CNNs permits us to use this astounding development. (ALVINN used a totally related organization which is very small by the current standard.)

While DAVE displayed the ability of beginning to end learning, and indeed was used to legitimize starting the DARPA Learning Applied to Ground Robots program [14], DAVE's show was not sufficiently reliable to give a full choice rather than more separated approaches to manage unpleasant landscape driving. DAVE's mean distance between crashes was around 20 m in complex conditions.

A couple of years earlier, another work was begun at NVIDIA that endeavored to foster DAVE and make a strong framework for driving on open streets. The central inspiration for this work is to keep away from the need to communicate human-given highlights, like way markings, screen rails, or different vehicles, and to put forth an attempt not required to make a gathering of "tolerating, then, at that point, else" rules, thinking about the perspective on these provisions. This chapter portrays the groundwork outcomes of this new application.

2.5 DAVE-2

The game plan framework for the getting-ready data for DAVE-2 has three cameras that were mounted behind the windshield (the dashboard) of the information securing vehicle. The time-wandered recordings from the cameras are acquired in the interim the controlling point applied by the human driver. This immediate solicitation is obtained by taking advantage of the vehicle's controller area network transport to make the framework free of the vehicle. This tends to the organizing request as 1/a where "a" is the turning range in meters. We utilize 1/a rather than "a" to frustrate an eccentricity when driving straight (the turning range for driving straight is unbelievable). 1/a successfully moves through zero from left turns (negative attributes) to right turns (positive qualities).

Getting ready data contains single images researched from the video, composed with the differentiating planning demand (1/a). Arranging with information from basically the human driver is not good. The affiliation should figure out some approach to improve from bungles. In any case, the vehicle will ceaselessly drift off the street. The preparation data is consequently extended with extra photographs that show the vehicle in various improvements from the mark of union of the way and rebellions from the track of the road.

Images for two unequivocal disproportionate advancements can be obtained from the left and right cameras. Extra moves between the cameras and all turns are reflected by a perspective difference in the picture from the closest camera. Cautious perspective change requires 3D scene information which we do not have. We along these lines reasoned the change by expecting that all bright lights under the skyline are on level ground and all focuses into the great beyond are massively far away. This winds up mind-blowing for a level region in any case it presents mutilations for objects that stick over the ground, like vehicles, shafts, trees, and developments. Luckily, these turns do not resolve a huge issue for network arranging. The controlling engraving for changed pictures is acclimated to the one that would coordinate the vehicle back to the best district and course in 2 s.

Images are managed into a CNN which then, at that point, cycles a proposed coordinating solicitation. The proposed demand veers from the ideal solicitation for that image, and the stacks of the CNN are changed in accordance with the CNN yield nearer to the best yield. The weight change is refined by utilizing the back spread as done in the Torch 7 AI store of our plan framework.

2.6 Network architecture

Exactly when we say CNN, generally we insinuate a 2D CNN which is used for picture gathering. However, there are two various types of CNNs used in all actuality, which are 1D and 3D CNNs. In this associate, we will cover 1D and 3D CNNs and their applications in actuality. I am expecting you are currently familiar with the possibility of convolution networks, in general. The 2D layer makes a convolution part that is convolved with the layer commitment to convey a tensor of yields. 2D and 3D convolution layers were expected to perform feature extraction and were picked observationally through a movement of examinations that changed layer plans.

2.6.1 Conv2D

This standard CNN [16] was initially presented within Lenet-5 engineering. It is for the most part utilized in icon information. On grounds that the part slides along two measurements on the information, it is termed as 2D CNN. The whole advantage of using CNN is that it can separate the spatial features from the data using its pieces, which various associations cannot do. For instance, CNN can identify edges, dispersion of shadings, and so forth in the picture which makes these organizations exceptionally strong in picture characterization and other comparable information that contains spatial properties.

2.6.2 Conv3D

Conv3D is generally used with 3D picture information. This layer makes a convolution part that is convolved with the layer obligation to pass on a tensor of yields. Like magnetic resonance imaging information, X-bar information is broadly utilized for examining the cerebrum, spinal strings, internal organs, and some more. A computerized tomography scan is comparatively an outline of 3D information, which is made by joining a development of X-transmit pictures taken from various concentrations around the body. We can utilize Conv3D to depict this clinical information or concentrate highlights from it.

The mean squared mistake is decreased in the heaviness of the organizations which exist between the directing order yield by the organization and the request for either the human driver, or the changed controlling request for awry and turned pictures. Early on layer of the affiliation performs picture standardization. The normalizer is hard-coded and isn't changed in the learning affiliation. Performing standardization in the affiliation permits the standardization to be changed with the affiliation plan and to be sped up through GPU preparing.

2.6.3 ReLU activation function

Activation is fundamentally a straightforward capacity that changes its contributions to yields that have a specific reach. Initiation function is otherwise called transfer work. It is used to choose the yield of neural association like yes or no. It maps the resulting characteristics in 0 to 1 or −1 to 1, and so on. Two types exist basically as linear and nonlinear. ReLU (rectified linear unit) is the most utilized initiation work right now [17].

$R(x) = \max(0,x)$ in which 0 is returned for negative input and x for positive. Both the function and its derivative are monotonic.

ReLU consists of no hefty calculation as there is no confounded math. So the model can set aside less effort to prepare or run. One more significant property to consider the benefit of utilizing ReLU is sparsity. It has turned into the default work for such neural learning because a model that uses it is easier to get ready and much of the time achieves better execution.

2.6.4 Max pooling

Pooling is an important feature taken into CNN models. The fundamental thought behind a pooling layer is to "gather" highlights from maps produced by convolving a channel over a picture. Authoritatively, its ability is to progressively diminish the spatial size of the depiction to decrease the proportion of limits and computation in the association [18]. The most widely recognized type of pooling is max pooling.

Max pooling is never actually a part to help overfitting by giving a separated kind of the depiction. As well, it reduces the computational cost by diminishing the amount of limits to learn and gives basic translation invariance to the inside depiction. Max pooling is done by applying a greatest channel to (generally) noncovering subregions of the underlying portrayal. Different types of pooling are normal and general.

2.6.5 Flatten

When the pooled highlighted map is acquired, the subsequent third stage is to level it. Flattening includes changing the whole pooled highlight map lattice into a single column or segment which is then taken care of to the neural networks for further processing.

2.6.6 Dropout layer

The dropout layer is a cover that discredits the responsibility of specific neurons toward the accompanying layer and leaves unmodified all others. We can apply a dropout layer to the information vector, in which case it discredits a part of its features; yet we can moreover apply it to a mysterious layer, in which case it refutes some mysterious neurons. They are significant in preparing CNNs in light of the fact that they forestall overfitting on the preparation information. On the off chance that they are absent, the main cluster of preparing tests impacts the learning in a lopsided highway. The Table 1 shows the layers of the proposed Convolutional Neural network model.

Table 1: Layers of convolutional neural network model.

Model: sequential		
Layer (type)	**Output shape**	**Param. no.**
Lambda (lambda)	(None, 100, 100, 1)	0
Conv2d (Conv2D)	(None, 100, 100, 32)	832
Activation (activation)	(None, 100, 100, 32)	0
Max_pooling2d (Maxpooling2D)	(None, 50, 50, 32)	0
Conv2d_1 (Conv2D)	(None, 50, 50, 32)	25,632

Table 1 (continued)

Model: sequential		
Layer (type)	**Output shape**	**Param. no.**
Activation_1 (activation)	(None, 50, 50, 32)	0
Max_pooling2d_1 (Maxpooling2D)	(None, 25, 25, 32)	0
Conv2d_2 (Conv2D)	(None, 25, 25, 64)	51,264
Activation_2 (activation)	(None, 25, 25, 64)	0
Max_pooling2d_2 (Maxpooling2D)	(None, 12, 12, 64)	0
Conv2d_3 (Conv2D)	(None, 12, 12, 64)	36,928
Activation_3 (activation)	(None, 12, 12, 64)	0
Max_pooling2d_3 (Maxpooling2D)	(None, 6, 6, 128)	0
Conv2d_4 (Conv2D)	(None, 6, 6, 128)	73,856
Activation_4 (activation)	(None, 6, 6, 128)	0
Max_pooling2d_4 (Maxpooling2D)	(None, 3, 3, 128)	0
Conv2d_5 (Conv2D)	(None, 3, 3, 128)	147,584
Activation_5 (activation)	(None, 3, 3, 128)	0
Max_pooling2d_5 (Maxpooling2D)	(None, 1, 1, 128)	0
Flatten (flatten)	(None, 128)	0
Dropout (dropout)	(None, 128)	0
Dense (dense)	(None, 1024)	132,096
Dense_1 (dense)	(None, 256)	262,400
Dense_2 (dense)	(None, 64)	16,448
Dense_3 (dense)	(None, 1)	65
Total params: 747,105 Trainable params: 747,105 Nontrainable params: 0		
None A total of nearly 7.47 lakh parameters		

3 Data validation and testing

The data is separated into train/test/dev. The train data is used to train the model. The validation is used to check if the model is going in the constructive path toward global optimum by verifying with the test data while training, and dev data is used to fine-tune the model. The model is validated on data that is not used for training so that the model does not overfit the data. Figure 1 shows the steering angle detection by the hybrid deep learning model.

Figure 1: Steering angle detection by the hybrid deep learning model.

4 Simulation – sanity check

The live feed of a camera is captured and it is converted into images. The image is parsed through the CNN model we have trained. These images are parsed at each individual frame level. The output, that is, the steering angle, is used to rotate the steering wheel image using the cv2 functions. Each output is printed to the prompt screen to understand the output of the model. The live video is played so that we can make sense that the model actually works.

5 Conclusion

CNNs can get to know the entire endeavor of way and road following without manual rot into a road or way really looking at distinguishing proof, semantic pondering, way orchestrating, and control. A humble amount of planning data from not by and large 100 h of driving was satisfactory to set up the vehicle to work in various conditions, on streets, neighborhood, and private roads in brilliant, cloudy, and blustery conditions. The CNN can take in huge road features from an incredibly insufficient getting ready sign (directing alone). The structure learns, for example, to perceive the graph of a road without the need of expressed names during readiness.

More works relied on to chip away at the generosity of the association, to find systems to affirm the strength, and to additionally foster impression of the association's internal taking care of steps. A lot of simulation should be done if the approach is selected for real time since the domain includes human health.

References

[1] Bechtel MG, McEllhiney E, Kim M, Yun H. DeepPicar: A low-cost deep neural network-based autonomous car. In *2018 IEEE 24th International Conference on Embedded and Real-Time Computing Systems and Applications (RTCSA)*. IEEE, 11–21, 2018.

[2] Daily M, Medasani S, Behringer R, Trivedi M. Self-driving cars. *Computer*, 50(12):18–23, 2017.

[3] Bojarski M, Del Testa D, Dworakowski D, Firner B, Flepp B, Goyal P, Jackel LD et al. End to end learning for self-driving cars. *arXiv preprint arXiv:1604.07316*, 2016.

[4] Badue C, Guidolini R, Carneiro RV, Azevedo P, Cardoso VB, Forechi A, . . . De Souza AF. Self-driving cars: A survey. *Expert Systems with Applications*, 165:113816, 2021.

[5] Urmson C. Self-driving cars and the urban challenge. *IEEE Intelligent Systems*, 23(2):66–68, 2008.

[6] Rao Q, Frtunikj J. Deep learning for self-driving cars: Chances and challenges. In *Proceedings of the 1st International Workshop on Software Engineering for AI in Autonomous Systems*. 35–38, 2018.

[7] https://github.com/ndrplz/self-driving-car.

[8] LeCun Y, Boser B, Denker JS, Henderson D, Howard RE, Hubbard W, Jackel LD. Backpropagation applied to handwritten zip code recognition. *Neural Computation*, 1(4): 541–551, 1989.

[9] Krizhevsky A, Sutskever I, Hinton GE. ImageNet classification with deep convolutional neural networks. *Advances in Neural Information Processing Systems*, 25:1097–1105, 2012.

[10] Jackel LD, Sharman D, Stenard CE, Strom BI, Zuckert D. Optical character recognition for self-service banking. *AT&T Technical Journal*, 74(4):16–24, 1995.

[11] Berg A, Deng J, Fei-Fei L. Large scale visual recognition challenge (ILSVRC), 2010. *URL* http://www.image-net.org/challenges/LSVRC, *3*, 2010.

[12] Rudin C. Stop explaining black box machine learning models for high stakes decisions and use interpretable models instead. *Nature Machine Intelligence*, 1(5):206–215, 2019.

[13] Jackel LD, Krotkov E, Perschbacher M, Pippine J, Sullivan C. The DARPA LAGR program: Goals, challenges, methodology, and phase I results. *Journal of Field Robotics*, 23(11–12):945–973, 2006.

[14] Pomerleau DA. *Alvinn: An autonomous land vehicle in a neural network*. Carnegie-Mellon Univ Pittsburgh PA Artificial Intelligence and Psychology Project, 1989.

[15] Wang D, Qi F. Trajectory planning for a four-wheel-steering vehicle. In *Proceedings 2001 ICRA. IEEE International Conference on Robotics and Automation (Cat. No. 01CH37164)*. Vol. 4, IEEE, 3320–3325, 2001.

[16] Verma S. Understanding 1D and 3D Convolution Neural Network – Keras, 2019.

[17] Brownlee J. A gentle introduction to the rectified linear unit (ReLU). *Machine Learning Mastery*, 6:2019.

[18] Yu D, Wang H, Chen P, Wei Z . Mixed pooling for convolutional neural networks. In *International Conference on Rough Sets and Knowledge Technology*. Springer, Cham, 364–375, 2014.

Badr Hirchoua, Saloua El Motaki

Semantic segmentation with transfer learning for self-driving cars

Abstract: Self-driving cars and autonomous robotic systems rely on the precise perception and understanding of the surrounding environment. Semantic segmentation is a crucial component of road scene perception that offers semantic information. Recently, numerous approaches have been proposed for semantic segmentation. Despite their enhanced performance, these models need a large dataset for training, and they are afflicted by high computation complexity. Fine-tuning deep pretrained models has shown state-of-the-art performances. To this end, we present novel and functional deep fully convolutional neural networks (FCN) architectures based on transfer learning for semantic pixel-wise segmentation. The first one is a fine-tuned FCN-based VGG16, where the second uses this first model as an encoder and concatenates its output with a fine-tune-based Xception model, which works as a decoder. The second model breaks an input image into different coherent semantically meaningful regions. These regions are then classified into predetermined classes. Exhaustive experiments are executed on three real-world datasets that confirm the effectiveness of our models trained on large or small datasets. Our models have significantly improved road segmentation for self-driving cars compared to the state-of-the-art methods.

Keywords: semantic segmentation, deep learning, autonomous driving, self-driving cars, transfer learning

1 Introduction

A solid semantic understanding of the surrounding environment is crucial for self-driving. Semantic segmentation (SS) is among the fundamental tasks in self-driving systems. SS is a core computer vision, which has proved tremendous progress due to the evolution of deep learning (DL) [1]. It creates one of the finest grained visual scene understanding tasks through predicting solid semantic classifications for an input image. State-of-the-art SS models sit their difficulties and shortcomings, as their remarkable accuracy is achieved at a prohibitive cost of extreme resources,

Badr Hirchoua, National Higher School of Arts and Crafts (ENSAM), Hassan II University, Casablanca, Morocco, e-mail: badr.hirchoua@gmail.com
Saloua El Motaki, Computer Science Department, Faculty of Science Dhar El Mahraz, University Sidi Mohamed Ben Abdellah, Fez, Morocco, e-mail: saloua.elmotaki@usmba.ac.ma

https://doi.org/10.1515/9783110760477-005

both computational and memory requirements. Significantly, the segmentation output of every pixel quality requests high-resolution feature maps that are conserved over the network. In addition, it maintains a substantial perceptive field on the output [2]. Therefore, the produced network architectures contain many layers and substitute feature-volume downsampling, often with expanded convolutions of the rising rate [3]. This behavior pushes to meaningful workload concentration more profound in the network that occurs in latency-intensive inference.

The design of convolutional deep neural networks (DNN) has improved SS. Notably, the proposed systems are based on fully convolutional architectures [4]. Fundamental researches proposed atrous convolution [5,6] and encoder–decoder structures using skip connections [7] to utilize context and increase localization, respectively. Additional researches concentrated on real-time performance [8], leveraging various modalities including depth [9], and determining neighborhood-based supervision [10] for segmentation.

The success of DNN for SS profoundly relies on large-scale labeled datasets. Practically, it is hard to obtain such data. Transfer learning (TL) is a method to transfer knowledge learned for a given task to succeed in a related one. Hence, TL enables the reuse of pretrained models on a different but related task. TL is beneficial in more accelerated training and offers more reliable and stable results on relatively small datasets that require a large amount of data. The TL-based model uses entirely pretrained models or trains a source model on a large amount of available data, and uses this source model on a target task. In both cases, we can reuse the entire model or just a part of it.

Learning and extracting essential and deeply discriminative features from real-world input images are challenging tasks. However, TL is primarily employed to extract effective and deeply discriminative features. Moreover, image classification-based TL has shown impressive results. In DL, CNN-based TL is adopted for training on small datasets. Pretrained models have learned to extract a wide variety of features through training on large benchmark datasets. It is fundamental to create a robust architecture that is strong and generalizable across different scenes. Pretrained ImageNet [11] models have significantly enhanced the performance of CNN architectures. SS is a complex task comparing with classification. It necessitates predicting each single-pixel class instead of predicting the whole input discrete classes. In addition, the segmentation mechanism requires determining the localization of objects in the input image. Motivated by this, we propose two architectures that use the pretrained VGG16 and the Xception CNN networks. The importance-aware classification/segmentation aims to determine essential classes. Cars are more important than roads and sidewalks, where the sky is the least significant. Hence, misclassifying cars will introduce a severer punishment. Therefore, we consider two possible scenarios. In the first one, we investigate the usefulness of different DNN architectures for SS. Precisely, given an image input, the model predicts the associated mask. Note that in our work, we focus on predicting the road mask. On the other hand, the second scenario attempts to respond to the

image's pixels labeling for SS. In other words, the aim is to label pixels in images to determine meaningful, coherent areas. In the first model, we fine-tuned the VGG16, adding 6 new layers with a direct connection of layers 10, 14, and 18 from the original pretrained model. This model takes an image as input and predicts its mask. The second model uses the first one as an encoder and a fine-tuned Xception model as a decoder. Next, we concatenate the two models' outputs and pass the result into the last layer for pixel labeling. To summarize, our principal contributions are as follows:
- Fine-tuning VGG16 with a crucial modification of the last layers for mask prediction
- Studying two possible challenging scenarios of SS
- Using the first fine-tuned model for pixel labeling that is more challenging than predicting an image mask
- Fine-tuning the Xception model for the second task of pixel labeling, using the first model as the encoder
- Achieving very promising results on three real-world datasets

The chapter is arranged in five sections. Section 2 presents a summary of the related work in the field of image segmentation and TL. In Section 3, we illustrate the proposed models. Section 4 details the experiments and presents the results achieved from the evaluation scenarios. The results and achievements discussion are given in the same section. Finally, we summarize our contributions and present future work in Section 5.

2 Related works

This section briefly introduces some of the research works that have discussed the same problems examined in this work. Also, these models involve many hypotheses. These hypotheses, in turn, consist of several assumptions that are shortly presented in this chapter.

State-of-the-art SS methods have overgrown over the last years, principally through the application of convolutional neural networks. Long et al. [4] introduced an FCN for end-to-end SS. They readjust well-known classification networks into FCN and fine-tune their learned representations to be transferred to the segmentation task. After that, this topic has attracted extra attention. Based on FCN, Ronneberger et al. [7] proposed the U-Net, which consists of expansive and contracting paths. The first path caught the context, where the second one enabled precise localization. In addition, this network added a skip connection to enhance the full spatial resolution recovery. Unlike FCN, SegNet [12] employed an encoder–decoder structure. Precisely, the architecture is composed of an encoder, a decoder, and a pixel-wise classification layer. Noh et al. [13] used a deconvolution network with a similar decoder path used by SegNet, except they chose deconvolution modules to perform upsampling operations. Lin et al. [14] introduced to combine mid- and high-level semantic features utilizing an encoder–decoder architecture. Chen et al. [5] proposed DeepLab to predict higher resolution with a network larger receptive field. DeepLab used dilated convolutions that injected holes into the convolutional filters rather than utilizing the conventional convolution with downsampling. Besides, the authors examined the impact of atrous convolutions, and spatial pyramid pooling, multigrid, various backbones, and various training extensions in DeepLab V3 [15] and DeepLab V3+ [16]. Despite their performances, these techniques demand sufficient pixel-level labeling for different classes while training. Besides, their generalizability to new classes with fewer labeled entries is weak.

Later, fundamental works on adaptive segmentation have been performed. Li et al. [17] fused neural architecture search with a trainable dynamic routing mechanism. Their framework generates data-dependent routes at run time. Nevertheless, by including the computation cost to the loss function, the framework requires more flexibility for customizing deployment, especially in varying requirements applications or over different devices without retraining.

Li et al. [18] proposed layer cascade (LC) and investigated one of the mechanisms used in our work, the early stopping for segmentation. Their work addresses segmentation as an extensive collection of independent classification tasks. Specifically, every pixel contributes only to the next exit if the newest prediction exceeds a confidence threshold, although, due to different per-pixel paths, this design pushed the system to massively unstructured computations. Moreover, LC established a handcrafted architecture, massively related to Inception-ResNet, and did not utilize several backbones or readjust its architecture to the objective device's capacities.

3 Method

In this section, we first formulate the SS task. Then, we highlight the used pre-trained models and detail the models' architecture.

3.1 Semantic segmentation task

An image with dimensions H, W height and width, respectively, can be described as $x \in [0,1]^{H \times W \times C}$, where C is the number of channels. Given the number of classes as N_c, the SS model returns $f(x) \in [0,1]^{H \times W \times N_c}$. This result denotes the predicted class probability corresponding to each pixel in the input image. Especially, $f_i^j(x) \in [0,1]$ expresses the predicted probability rate for the image pixel i associated with the class j. Therefore, the predicted SS $SS(x) \in {}^{H\times W}$ is calculated by selecting the classes with the largest probability rate in each image pixel:

$$SS(x) = \operatorname{argmax}_{j\in\{1,\ldots,N_C\}} f_i^j(x) \quad \forall i \in \{1, \ldots, H\times W\}$$

The SS x ground truth is described as $y \in \mathbb{N}^{H\times W}$ and attaches the true class (in $\{1, \ldots, N_c\}$) to each image pixel. The cross-entropy loss $L_{CE}(f_i(x), y_i) = -\log(f_i^{y_i}(x))$ is used to evaluate the SS model performance. Hence, for each image pixel i, the predicted value f_i is matched on the ground truth class y_i.

3.2 Pretrained models

For the AlexNet groundbreaking win on the competition of ImageNet, CNNs were (re-)raised as top techniques for resolving complicated computer vision tasks [19]. Therefore, powerfully enhanced architectures such as Xception [20] or VGG16 [21] have been developed. CNNs have demonstrated exceptional achievement on various computer vision tasks.

The VGG16 CNN comprises five convolutional blocks containing two or three convolutional layers. The number of filters in each block is the same, and it doubles after every max-pooling step. The network consists of 13 convolution layers with a single stride and applies 1 stacked padding with 3 fully connected layers. The softmax activates the last fully connected layer that is utilized for classification. The remaining layers use the rectified linear unit activation function. In our work, we have removed the network classification block and preserved the parameters of the original network. Xception is the ultimate Inception version, which is based on depthwise separable convolution layers. Precisely, these layers are a spatial convolution executed separately over every input channel, followed by a pointwise 1 × 1 convolution that adjusts the input dimensions. Xception is trained on the ImageNet dataset [11]. The Xception architecture is split into three major components: entry, middle, and exit flow. It contains 36 convolution layers that are arranged into 14 modules. All modules have linear residual connections, except the first and last ones.

3.3 Network architecture

Figure 1 highlights the first proposed TL model, which is based on VGG16 model trained on ImageNet for 1,000 classes. We freeze all VGG16 layers, and we delete the classification layers. Besides, we define the 10th layer output as block 3 output, the block 4 output as the 14th layer output, and the block 5 output as the 18th layer output. Next, the green layers represent each convolutional 1 × 1 layer to maintain spatial information of different defined layers (blocks 3, 4, and 5). Note that a deconvolutional layer follows each convolutional layer except the last layer. Finally, two skip connections are injected to secure the vanishing gradient problem. This DNN inputs an RGB image and outputs its associated segmentation. In this model, we focus only on on-road segmentation tasks without labeling.

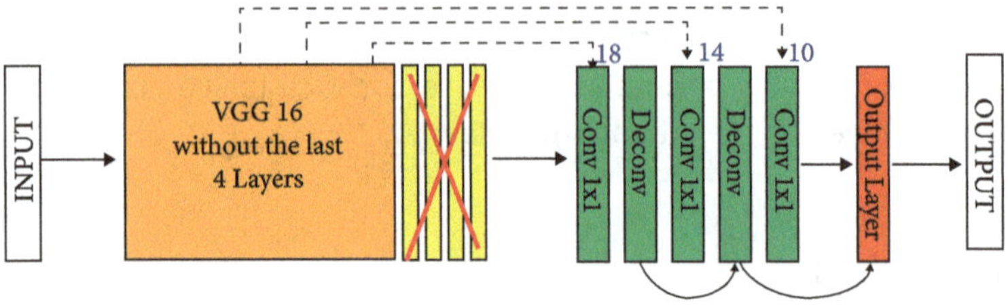

Figure 1: The block diagram of the transfer learning model-based VGG16.

On the other hand, the second model contains two DNNs. The first one is our first model, in which we delete the last layer only and add an output layer with the same size as the last layer. In this rectified network, the input image is fed to the network-based VGG16, which produces a predicted output 1. Next, we multiply the input image and the generated output 1, which is an input for the second modified network-based Xception that provides output 2. The Xception Net [22] has top five accuracy of 94.5%, which makes it a good base for TL. Finally, we concatenate outputs 1 and 2 and feed the final output to the sigmoid to generate the pixel label. Contrary to Ref. [23] that use double U-Net-based VGG19, our first model helps us to avoid fetching the input image and its corresponding output 1 again to the first network, meaning less computation time and resources. Note that the first model is trained on the same data without augmentation (Figure 2).

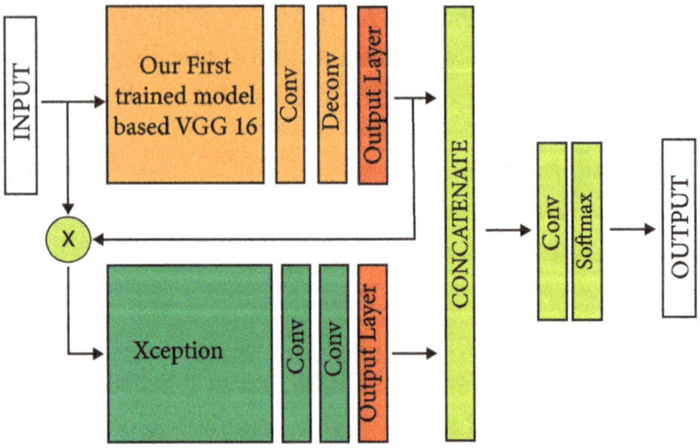

Figure 2: The block diagram of the transfer learning model based on our first model and the Xception network.

4 Experiments

In this section, we show the evaluations conducted to review the performance of our models in two possible scenarios.

4.1 Datasets

The experiments were taken out utilizing the Carla dataset [24]. It provides 5,000 data images and labeled SS taken via CARLA self-driving car simulator. In our context, the ground truth involves sets of RGB and the corresponding semantic segments. Figure 3 shows some samples from the dataset without any modifications. Because the dataset is relatively small, additional images are simulated using augmentation to increase the number of samples during training. Precisely, we have used the transformation matrix, shifts, random rotations, reflections, shearings, and zooms. They are carried out directly on the tensor processing units (TPUs) for each image and its associated mask. Figure 4 highlights four augmented samples. In addition to the Carla dataset, the CamVid [25] is adopted for evaluating the performance of self-driving systems. The images are taken from a driving perspective. The dataset implies 11 essential target classes that are usually experienced in a driving scenario. Since the CamVid includes 367 training samples, 26 validation samples, and 233 test samples, we include the Cityscapes [26] dataset in our experiments. Cityscapes is a semantic scene understanding dataset containing 2,975 training samples, 500 validation samples, and 1,525 test samples. The last dataset used in this work is the SemanticKITTI [27] dataset. SemanticKITTI is one of the largest datasets used on autonomous vehicle segmentation tasks. This dataset contains 41,000 total frames captured in 21 sequences.

4.2 Implementation scenarios

In order to respond to the SS challenge, we conduct two main experiments to cover two possible scenarios. The first scenario investigates the usefulness of different DNN architectures for SS. The set of experiments of this scenario intends to examine our method's performance for SS. Precisely, in this scenario, we mainly focus on road segmentation. To do so, in addition to our model, we have implemented the U-Net [7] and SegNet [12] encoder–decoder networks, run different experiments, and compared different results in terms of loss and accuracy. In these experiments, we have used the GPUs to accelerate the training.

The second scenario attempts to respond to the image's pixels labeling for SS. The set of experiments of this scenario aims to label pixels in images to determine meaningful, coherent areas. Figure 5 highlights the pixel number per label. The importance-aware methods [28] claim that the separation between object/pixel importance requires

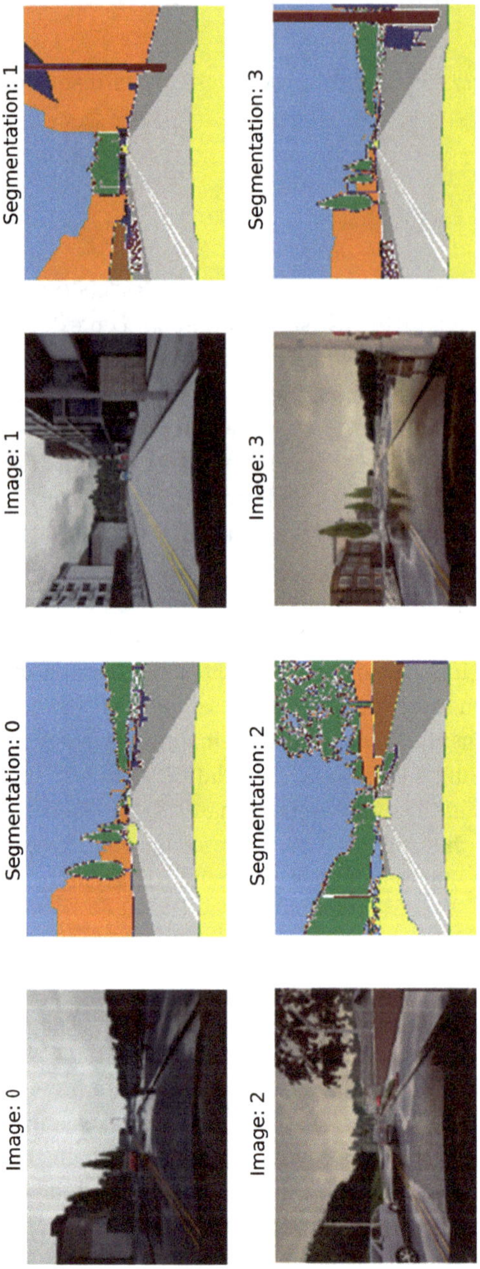

Figure 3: Samples of the training dataset.

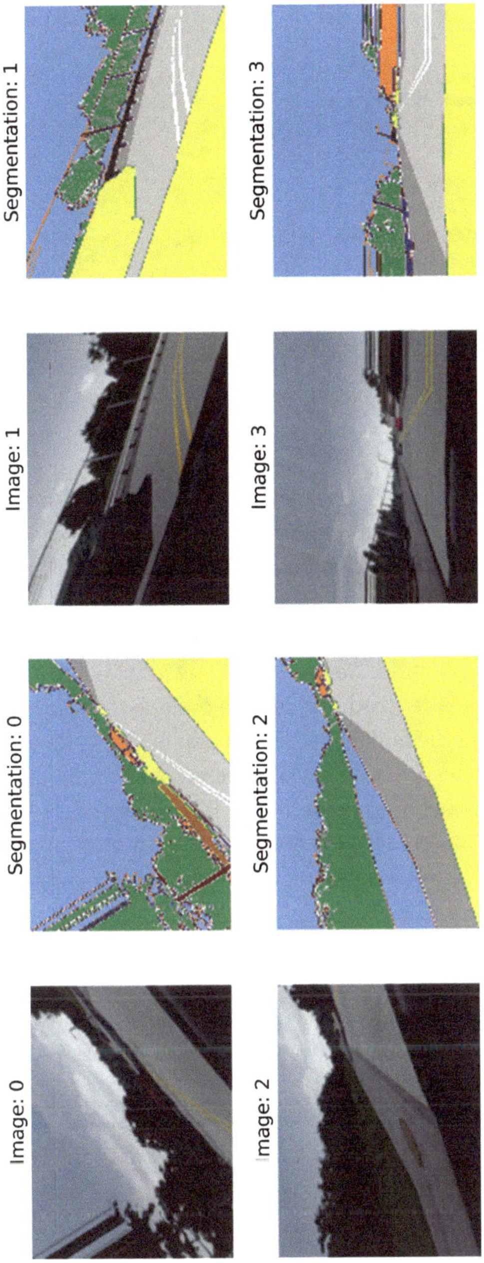

Figure 4: The additional images simulated with augmentation.

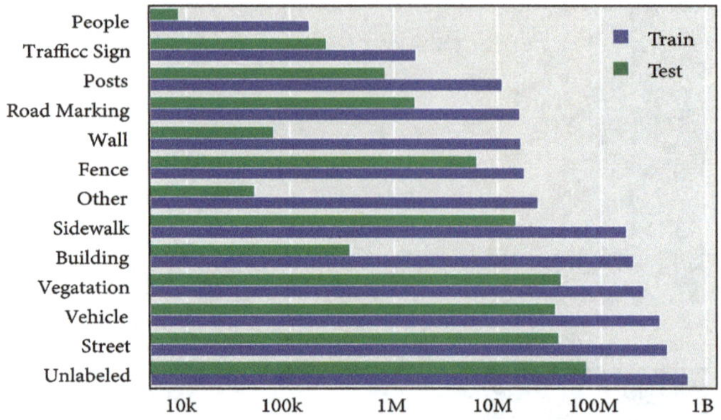

Figure 5: The pixel number per label.

to be exerted under attention. Therefore, we consider different rich classes chosen from the dataset: building, fence, pedestrian, pole, road line, road, sidewalk, vegetation, car, wall, and a traffic sign. Precisely, the number of singular pixel labels is split into three hierarchies of importance. This unbalanced combination can point to the wholly represented labels (unlabeled, vehicle, and street), crushing the gradients of the limited classes (traffic signs and people). In this scenario, *ReduceLROnPlateau* monitors the training process, and as soon as a plateau is reached, the learning rate is reduced. In addition, the model is accompanied by *EarlyStopping*, which is assumed to prevent overfitting. The early stopping [29] optimizes the number of epochs hyper-parameter directly. Practically, it ensures that the model avoids overfitting the training data by stopping the training procedure early. The compiler terminates the training if the model shows no improvement for a fixed value of consecutive epochs. In these experiments, we have used the TPUs that accelerate the training on the big augmented images enormously compared to CPUs and GPUs. Figure 4 shows four samples of the augmented images.

4.3 Training loss

To investigate the performance of different models, we utilize binary cross-entropy and the accuracy to evaluate the first scenario and sparse categorical cross-entropy alongside sparse categorical accuracy for the second scenario.

Cross-entropy [30]: This is the criterion applied in image segmentation addresses the minimal error of information content. It is determined as the difference between two probability distributions. It is broadly used for the classification task, and it works well in segmentation for pixel-level classification. Binary cross-entropy is given as follows:

$$L_{\mathrm{BCE}}(y, \hat{y}) = -(y \log(\hat{y}) + (1-y)\log(1-\hat{y})) \tag{1}$$

where $\hat{y}$ and y represent the predicted value and the ground truth, respectively.

Categorical cross-entropy: The categorical cross-entropy loss function uses equation (2) to calculate the loss of an example:

$$\mathrm{Loss} = -\sum_{i=1}^{k} y_i \cdot \log \hat{y}_i \tag{2}$$

where $\hat{y}_i$ is the ith model output, y_i represents the corresponding target value, and k is the value number in the model prediction.

This loss reflects the distinctiveness of two discrete probability distributions. Given this, y_i represents the probability that i occurs, where all y_i are up to 1. The minus sign guarantees that the loss becomes smaller during the closeness of distributions to each other.

4.4 Experimental results

Figure 6 shows the accuracy and loss performances for our models SegNet and U-Net, respectively. Each model is trained using Adam optimizer alongside the binary cross-entropy loss. The first model runs for 30 epochs, the second model for 35 epochs, and the last model for 50 epochs. The figure highlights the accuracy improvement through epochs and loss/error reduction with each epoch (i.e., advance through the input images). Clearly, all networks present slight deviations in the performance over epochs. Our model converges very quickly in a few epochs. All models record high accuracy with a slight improvement of our model over others. However, this first scenario does not demonstrate the superiority of our model; thus, in the second scenario, we have targeted a complex task. Precisely, we consider the image pixels labeling for SS. Figure 7 shows the visual results of the proposed model. We can observe that the segmentation produced by our model is highly identical to the ground truth.

Before proceeding to the second scenario experiments, we compare our second model that uses the first model to a model that uses VGG16 without modification, except deleting the last four layers. Figure 8 shows the comparisons in terms of sparse categorical loss and sparse categorical accuracy. Our model converges faster than the other one, which confirms our choice of architecture. The new layers added to the VGG16 used for TL are making deference. Precisely, our model outputs a dense representation that can be used for different purposes, including SS. Figure 9 shows the learning rate changes during model training. The *ReduceLROnPlateau* mechanism monitors the training process. This mechanism controls the loss function. In other words, if the loss does not decrease for a specific number of epochs (five in our model), then the learning rate is lowered. It starts at 0.001 and ends by 3.906 e-06.

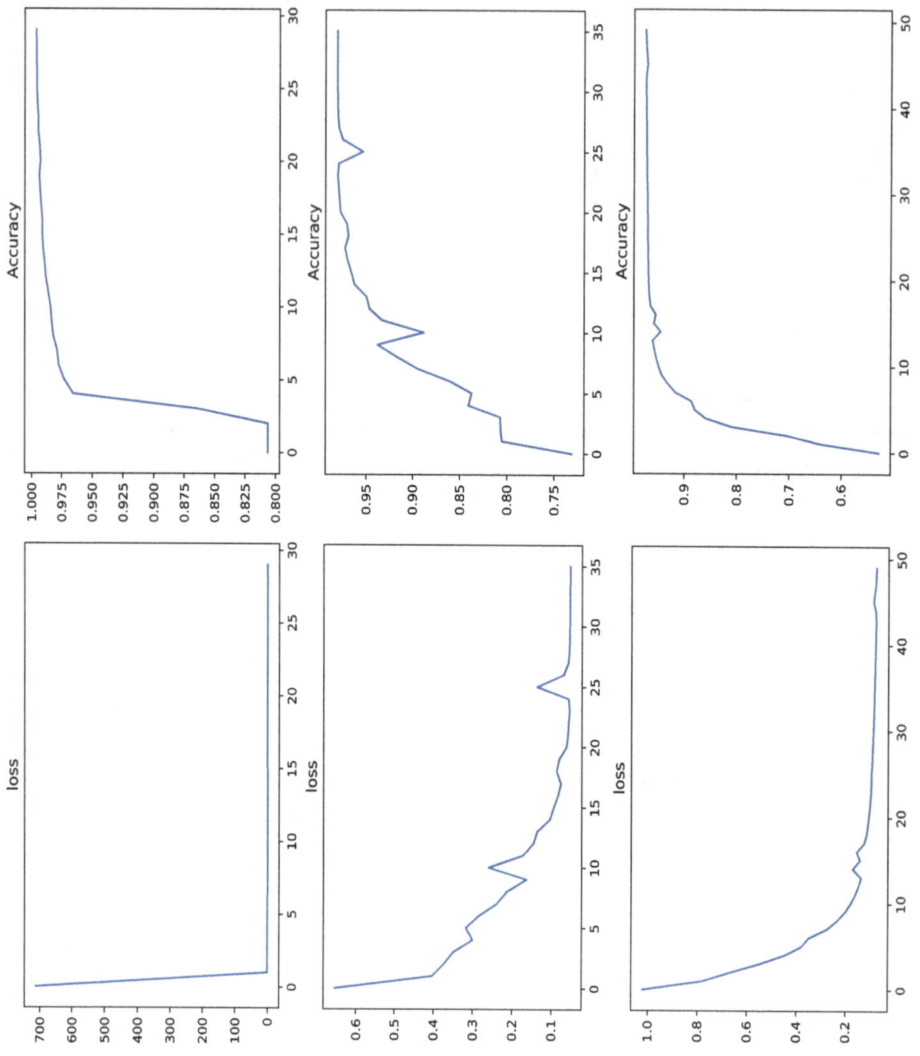

Figure 6: Our model results for scenario 1. First figure represents the accuracy during the training, where the second describes the loss during the training.

Using the Carla dataset, our model achieves the results highlighted in Figure 10, which represents the confusion matrix. Among other advantages, the confusion matrix can show which labels pose a challenge. Alongside Figure 5 and considering the context of this work, we are interested in vehicle and street. However, the label unbalance brings instability to the whole learning process though the two models predict the targeted label accurately. The models predict the vehicles that are truly vehicles with a 94% accuracy and streets with 86%.

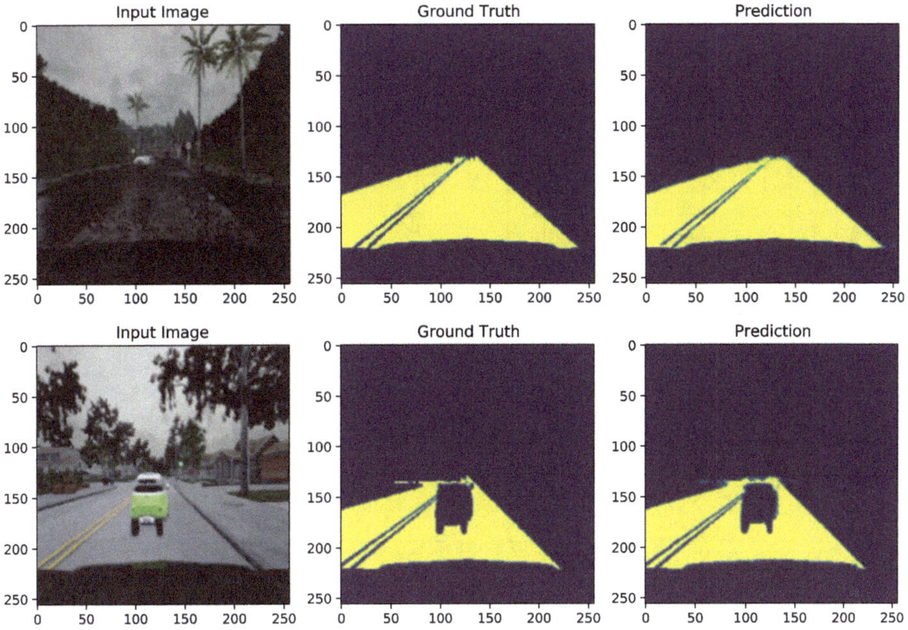

Figure 7: Our model results. First column is the: input image from dataset; second column is the predicted segmentation from our network; and third column is the ground truth segmentation.

To demonstrate the effectiveness of our model, we compare our model with three existing DNNs: FCN [26], SegNet [12], and ENet [31]. The ENet is a deep network for autonomous driving purposes. For CamVid and Cityscapes datasets, we use the Intersection over Union (IoU) scores on test sets to report different achievements. In these experiments, we consider the following classes: road, sidewalk, vehicle, and building. The results are shown in Tables 1 and 2, respectively. Results in both tables demonstrate that the considered classes are segmented with very high IoU scores by all models with superiority of our model. Specifically, the IoU scores of road, sidewalk, and vehicle generated by our model are the highest scores, reflecting the effectiveness of our model in SS related to automated driving.

The last experiments compare our model with the other published models. Table 3 presents the results and confirms that our model reaches the state-of-the-art performance in terms of mean IoU. Our model records a 1.5% increase from the second best method [45]. Our model dominates considerably in classifying different objects; thus, it is reliable for understating complicated scenes and environments.

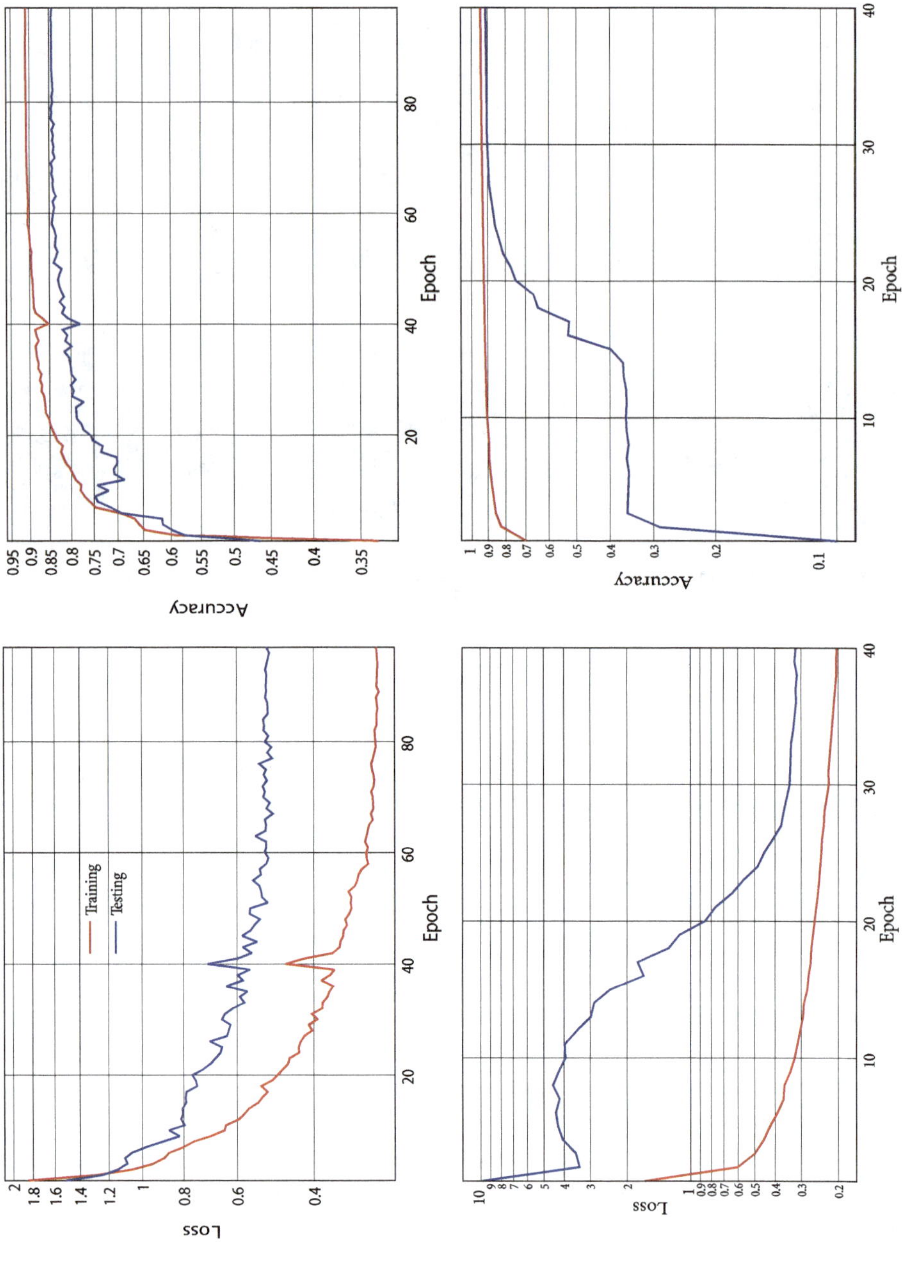

Figure 8: The first figure represents our second model that uses the first retrained model, where the second uses the VGG16 as encoder and the Xception Net as a decoder.

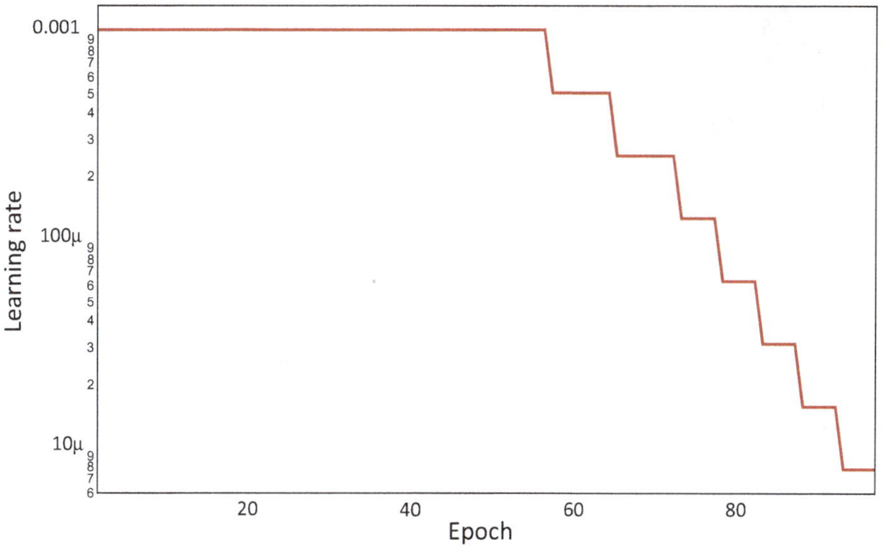

Figure 9: The learning rate during the training.

5 Conclusion

In this chapter, we have introduced two functional deep FCN architectures based on TL for semantic pixel-wise segmentation. Precisely, we have fine-tuned two models using two pretrained models: the VGG16 and the Xception. In the first model, we have replaced the last four layers of the VGG16 model with convolutional blocks. After training this model and evaluating its performance, we have used it as an encoder for the second model. Besides, its output is concatenated with fine-tuned Xception model to split the input image into different coherent semantically meaningful areas. We have conducted numerous experiments to examine the performance of our models alongside several state-of-the-art models, adopting several pretrained models. Also, we have investigated the convergence as well as the hyperparameters control using different mechanisms. The overall obtained results revealed that our models outperform the state-of-the-art model with a significant score. Finally, our models reached a highly promising accuracy and surpassed the evaluated model on three real-world datasets.

In future work, we plan to investigate the robustness of SS for self-driving cars, facing real-world adversarial perturbations. Moreover, we will further introduce solid resistance mechanisms to enhance the robustness of SS models.

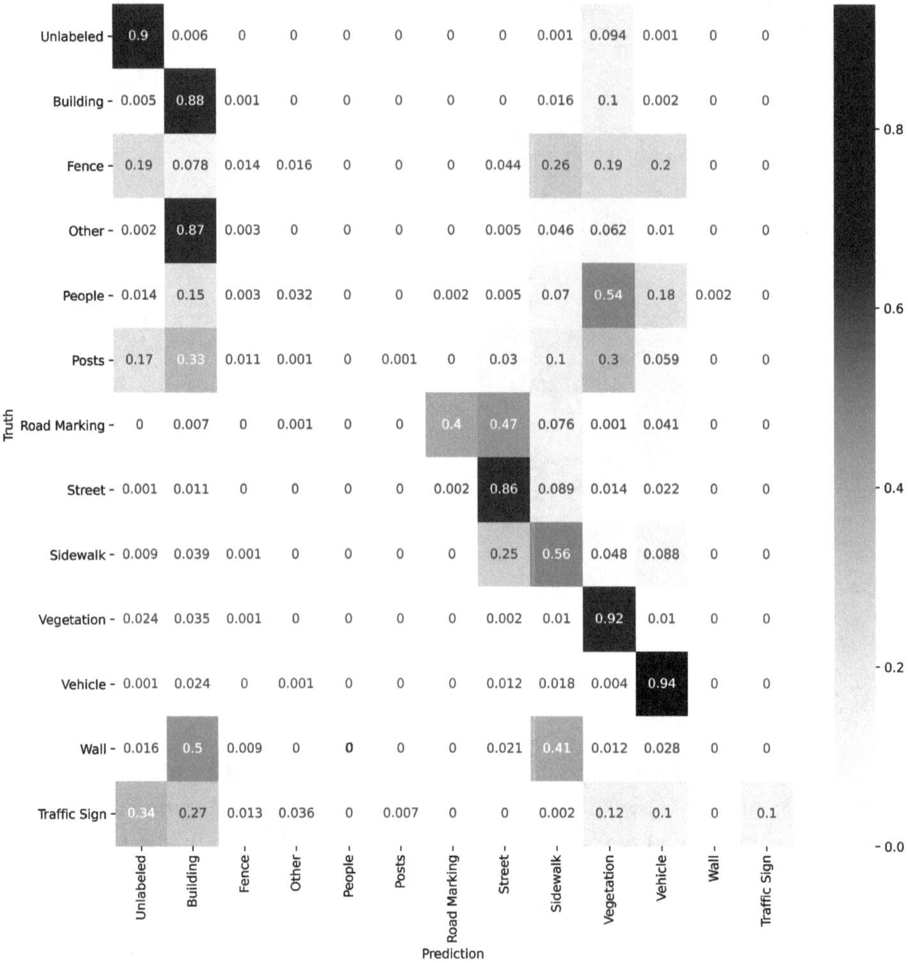

Figure 10: The confusion matrix with an overall accuracy of 84.5%.

Table 1: The comparison results of various methods using CamVid dataset.

	Road	**Sidewalk**	**Vehicle**	**Building**	**Mean IoU**
ENet	95.1	86.7	82.4	74.7	68.3
SegNet	97.2	84.4	82.1	88.8	65.2
FCN	98.1	89.5	84.5	**93.7**	69.7
Ours	**98.4**	**91.7**	**91.3**	92.6	**72.1**

Table 2: The comparison results of various methods using Cityscapes dataset.

	Road	Sidewalk	Vehicle	Building	Mean IoU
ENet	96.3	74.2	90.6	85.0	58.3
SegNet	96.4	73.2	89.3	84.0	57.0
FCN	97.3	77.6	91.9	**88.7**	64.3
Ours	**98.6**	**88.1**	**93.2**	88.2	**72.8**

The bold values are the highest ones, and therefore, they reflect the best performances.

Table 3: The comparison results of various methods using SemanticKITTI dataset.

	Road	Sidewalk	Vehicle	Building	Mean IoU
S-BKI [32]	92.6	77.4	83.8	89.7	51.3
RangeNet++ [33]	91.8	75.2	91.4	87.4	52.2
LatticeNet [34]	90.0	74.1	92.9	88.2	52.9
RandLA-Net [35]	90.7	73.7	94.2	86.9	53.9
PolarNet [36]	90.8	74.4	93.8	90.0	54.3
MinkNet42 [37]	91.1	69.7	94.3	**92.7**	54.3
3D-MiniNet [38]	91.6	74.5	90.5	89.4	55.8
SqueezeSegV3 [39]	91.7	74.8	92.5	89.0	55.9
Kpconv [40]	88.8	72.7	96.0	90.5	58.8
SalsaNext [41]	91.7	75.8	91.9	90.2	59.5
FusionNet [42]	91.8	77.1	95.3	92.5	61.3
KPRNet [43]	93.2	**80.6**	95.5	91.7	63.1
SPVNAS [44]	90.2	75.4	97.2	91.6	67.0
$(AF)^2$-S3Net [45]	91.3	72.5	94.5	87.9	69.7
Ours	**94.3**	78.2	**98.3**	89.6	71.2

The bold values are the highest ones, and therefore, they reflect the best performances.

References

[1] Ghosh S, Das N, Das I, Maulik U. Understanding deep learning techniques for image segmentation. *ACM Computing Surveys*, 52(4):2019.

[2] Lin T-Y, Maire M, Belongie S, Hays J, Perona P, Ramanan D, Dollár P, Lawrence Zitnick C. Microsoft coco: Common objects in context. In *European conference on computer vision*. Springer, 740–755, 2014.

[3] Yu F, Koltun V, Funkhouser T. Dilated residual networks. In *2017 IEEE Conference on Computer Vision and Pattern Recognition (CVPR)*. 636644, 2017.

[4] Long J, Shelhamer E, Darrell T. Fully convolutional networks for semantic segmentation. In *2015 IEEE Conference on Computer Vision and Pattern Recognition (CVPR)*. 3431–3440, 2015.

[5] Chen L-C, Papandreou G, Kokkinos I, Murphy K, Yuille AL. DeepLab: Semantic image segmentation with deep convolutional nets, atrous convolution, and fully connected CRFs. *IEEE Transactions on Pattern Analysis and Machine Intelligence*, 40(4):834–848, 2018.

[6] Yu F, Chen H, Wang X, Xian W, Chen Y, Liu F, Madhavan V, Darrell T. Bdd100k: A diverse driving dataset for heterogeneous multitask learning. In *2020 IEEE/CVF Conference on Computer Vision and Pattern Recognition (CVPR)*. 2633–2642, 2020.

[7] Ronneberger O, Fischer P, Brox T. U-net: Convolutional networks for biomedical image segmentation. In *International Conference on Medical Image Computing and Computer-Assisted Intervention*. Springer, 234–241, 2015.

[8] Yu C, Wang J, Peng C, Gao C, Yu G, Sang N. Bisenet: Bilateral segmentation network for real-time semantic segmentation. In *Proceedings of the European Conference on Computer Vision (ECCV)*. 325–341, 2018.

[9] Xu D, Ouyang W, Wang X, Sebe N. Pad-net: Multi-tasks guided prediction-and-distillation network for simultaneous depth estimation and scene parsing. In *2018 IEEE/CVF Conference on Computer Vision and Pattern Recognition*. 675–684, 2018.

[10] Ke T-W, Hwang J-J, Liu Z, Yu SX. Adaptive affinity fields for semantic segmentation. In *Proceedings of the European Conference on Computer Vision (ECCV)*. 587–602, 2018.

[11] Russakovsky O, Deng J, Su H, Krause J, Satheesh S, Ma S, Huang Z, Karpathy A, Khosla A, Bernstein M, Berg AC, Fei-Fei L. ImageNet large scale visual recognition challenge. *International Journal of Computer Vision*, 115(3):211–252, 2015.

[12] Badrinarayanan V, Kendall A, Cipolla R. Segnet: A deep convolutional encoder-decoder architecture for image segmentation. *IEEE Transactions on Pattern Analysis and Machine Intelligence*, 39(12):2481–2495, 2017.

[13] Noh H, Hong S, Han B. Learning deconvolution network for semantic segmentation. In *2015 IEEE International Conference on Computer Vision (ICCV)*. 1520–1528, 2015

[14] Lin G, Milan A, Shen C, Reid I. RefineNet: Multi-path refinement networks for high-resolution semantic segmentation. In *2017 IEEE Conference on Computer Vision and Pattern Recognition (CVPR)*. 5168–5177, 2017.

[15] Chen L-C, Papandreou G, Schroff F, Adam H. Rethinking atrous convolution for semantic image segmentation. *arXiv preprint arXiv:1706.05587*, 2017.

[16] Chen L-C, Zhu Y, Papandreou G, Schroff F, Adam H. Encoder-decoder with atrous separable convolution for semantic image segmentation. In *Proceedings of the European conference on computer vision (ECCV)*. 801–818, 2018.

[17] Li Y, Song L, Chen Y, Li Z, Zhang X, Wang X, Sun J. Learning dynamic routing for semantic segmentation. In *2020 IEEE/CVF Conference on Computer Vision and Pattern Recognition (CVPR)*. 8550–8559, 2020.

[18] Li X, Liu Z, Luo P, Loy CC, Tang X. Not all pixels are equal: Difficulty-aware semantic segmentation via deep layer cascade. In *2017 IEEE Conference on Computer Vision and Pattern Recognition (CVPR)*. 6459–6468, 2017.
[19] LeCun Y, Bengio Y, Hinton G. Deep learning. *Nature*, 521(7553):436–444, 2015.
[20] Chollet F. Xception: Deep learning with depthwise separable convolutions. In *2017 IEEE Conference on Computer Vision and Pattern Recognition (CVPR)*. 1800–1807, 2017.
[21] Simonyan K, Zisserman A. Very deep convolutional networks for large-scale image recognition. *arXiv preprint arXiv:1409.1556*, 2014.
[22] Chollet F. Xception: Deep learning with depthwise separable convolutions. In *Proceedings of the IEEE conference on computer vision and pattern recognition*. 1251–1258, 2017.
[23] Jha D, Riegler MA, Johansen D, Halvorsen P, Johansen HD. Doubleu-net: A deep convolutional neural network for medical image segmentation. In *2020 IEEE 33rd International Symposium on Computer-Based Medical Systems (CBMS)*. 558–564, 2020.
[24] Dosovitskiy A, Ros G, Codevilla F, Lopez A, Koltun V. CARLA: An open urban driving simulator. In *Proceedings of the 1st Annual Conference on Robot Learning*. 1–16, 2017.
[25] Brostow GJ, Shotton J, Fauqueur J, Cipolla R. Segmentation and recognition using structure from motion point clouds. In Forsyth D, Torr P, Zisserman A editors, *Computer Vision – ECCV 2008*. Springer, Berlin Heidelberg, Berlin, Heidelberg, 44–57, 2008.
[26] Cordts M, Omran M, Ramos S, Rehfeld T, Enzweiler M, Benenson R, Franke U, Roth S, Schiele B. The cityscapes dataset for semantic urban scene understanding. In *Proceedings of the IEEE Conference on Computer Vision and Pattern Recognition*. 3213–3223, 2016.
[27] Behley J, Garbade M, Milioto A, Quenzel J, Behnke S, Stachniss C, Gall J. SemanticKITTI: A dataset for semantic scene understanding of LiDAR sequences. In *2019 IEEE/CVF International Conference on Computer Vision (ICCV)*. 9296–9306, 2019.
[28] Chen B, Gong C, Yang J. Importance-aware semantic segmentation for autonomous vehicles. *IEEE Transactions on Intelligent Transportation Systems*, 20(1):137–148, 2019.
[29] Bengio Y. *Practical Recommendations for Gradient-Based Training of Deep Architectures*. Springer, Berlin Heidelberg, Berlin, Heidelberg, 437–478, 2012.
[30] Yi-de M, Qing L, Zhi-bai Q. Automated image segmentation using improved PCNN model based on cross-entropy. In *Proceedings of 2004 International Symposium on Intelligent Multimedia, Video and Speech Processing, 2004*. 743–746, 2004.
[31] Paszke A, Chaurasia A, Kim S, Culurciello E. ENet: A deep neural network architecture for real-time semantic segmentation. *arXiv preprint arXiv:1606.02147*, 2016.
[32] Gan L, Zhang R, Grizzle JW, Eustice RM, Ghaffari M. Bayesian spatial kernel smoothing for scalable dense semantic mapping. *IEEE Robotics and Automation Letters*, 5(2):790–797, 2020.
[33] Milioto A, Vizzo I, Behley J, Stachniss C. Rangenet ++: Fast and accurate LiDAR semantic segmentation. In *2019 IEEE/RSJ International Conference on Intelligent Robots and Systems (IROS)*. 4213–4220, 2019.
[34] Rosu RA, Schütt P, Quenzel J, Behnke S. LatticeNet: Fast point cloud segmentation using permutohedral lattices. *arXiv preprint arXiv:1912.05905*, 2019.
[35] Hu Q, Yang B, Xie L, Rosa S, Guo Y, Wang Z, Trigoni N, Markham A. Randla-net: Efficient semantic segmentation of large-scale point clouds. In *2020 IEEE/CVF Conference on Computer Vision and Pattern Recognition (CVPR)*. 11105–11114, 2020.
[36] Zhang Y, Zhou Z, David P, Yue X, Xi Z, Gong B, Foroosh H. Polarnet: An improved grid representation for online LiDAR point clouds semantic segmentation. In *2020 IEEE/CVF Conference on Computer Vision and Pattern Recognition (CVPR)*. 9598–9607, 2020.

[37] Choy C, Gwak J, Savarese S. 4D spatio-temporal ConvNets: Minkowski convolutional neural networks. In *2019 IEEE/CVF Conference on Computer Vision and Pattern Recognition (CVPR)*. 3070–3079, 2019.

[38] Alonso I, Riazuelo L, Montesano L, Murillo AC. 3d-mininet: Learning a 2D representation from point clouds for fast and efficient 3D LiDAR semantic segmentation. *IEEE Robotics and Automation Letters*, 5(4):5432–5439, 2020.

[39] Xu C, Wu B, Wang Z, Zhan W, Vajda P, Keutzer K, Tomizuka M. Squeezesegv3: Spatially-adaptive convolution for efficient point-cloud segmentation. In Vedaldi A, Bischof H, Brox T, Frahm J-M editors, *Computer Vision – ECCV 2020*, Springer International Publishing, Cham, 1–19, 2020.

[40] Thomas H, Qi CR, Deschaud J-E, Marcotegui B, Goulette F, Guibas L. KPConv: Flexible and deformable convolution for point clouds. In *2019 IEEE/CVF International Conference on Computer Vision (ICCV)*. 6410–6419, 2019.

[41] Aksoy EE, Baci S, Cavdar S. Salsanet: Fast road and vehicle segmentation in LiDAR point clouds for autonomous driving. In *2020 IEEE Intelligent Vehicles Symposium (IV)*. 926–932, 2020.

[42] Zhang F, Fang J, Wah B, Torr P. Deep fusionnet for point cloud semantic segmentation. In *Computer Vision–ECCV 2020: 16th European Conference, Glasgow, UK, August 23–28,2020, Proceedings, Part XXIV 16*. Springer, 644–663, 2020.

[43] Kochanov D, Nejadasl FK, Booij O. Kprnet: Improving projection-based LiDAR semantic segmentation. *arXiv preprint arXiv:2007.12668*, 2020.

[44] Tang H, Liu Z, Zhao S, Lin Y, Lin J, Wang H, Han S. Searching efficient 3d architectures with sparse point-voxel convolution. In *European Conference on Computer Vision*. Springer, 685–702, 2020.

[45] Cheng R, Razani R, Taghavi E, Enxu L, Liu B. 2-s3net: Attentive feature fusion with adaptive feature selection for sparse semantic segmentation network. In *Proceedings of the IEEE/CVF Conference on Computer Vision and Pattern Recognition*. 12547–12556, 2021.

Saloua El Motaki, Badr Hirchoua

An improved anomaly detection process in smart vehicle using a supervised ensemble-learning framework

Abstract: As the complexity of autonomous transportation systems increases with advances in connected and automated vehicle (CAV) technology, they are increasingly suffering from performance variations due to software- and hardware-related issues. Such problems may result in breakdowns and inefficiencies and are part of the critical challenges of system resilience. Reducing performance degradation requires timely and reliable detection and diagnosis of the anomalies causing the degradation, along with mitigating actions. However, identifying anomalies from the large, highly dimensional, and noisy data collected by system monitoring infrastructures is problematic. This chapter presents a machine learning–based framework for automatically detecting and diagnosing anomalous behavior in vehicles. Our model leverages historical data collected from various vehicle sensors to extract signatures of previously observed anomalies. By selecting the required features, our framework identifies anomalies and extracts their profiles. We evaluate the proposed model using simulations and demonstrate that it successfully identifies 98% of the injected anomalies.

Keywords: anomaly detection, performance degradation, automated vehicle, machine learning

1 Introduction

Autonomous vehicles are currently promoted by the media as ready to be commercialized (as early as 2022). However, the recent incidents and accidents of experimental autonomous vehicles circulating in the public environment (Tesla, Uber, etc.) have shown that operational safety has not been sufficiently addressed and that no satisfactory solutions have yet been found. The complexity of autonomous vehicles, of the driving tasks, and the open and uncertain environment in which they operate constitutes a challenge to apply dependability techniques [1].

Saloua El Motaki, Computer Science Department, Faculty of Science Dhar El Mahraz, University Sidi Mohamed Ben Abdellah, Fez, Morocco, e-mail: saloua.elmotaki@usmba.ac.ma
Badr Hirchoua, National Higher School of Arts and Crafts (ENSAM), Hassan II University, Casablanca, Morocco, e-mail: badr.hirchoua@gmail.com

https://doi.org/10.1515/9783110760477-006

In addition, the failures (including malfunctions linked to the degradation of components, improper installation, and potential implementation of inappropriate controls) of these vehicles expose passengers and bystanders to severe accidents. Environmental damage and economic losses are also to be expected. Autonomous vehicles are thus critical systems, which require reducing the risks of their malfunction.

Several mechanisms are built to ensure compliance with safety concerns (antivirus, firewall, etc.); however, we cannot guarantee to cover all possible threats, especially in sophisticated environments such as self-driving cars. The anomaly detection mechanism has been introduced in the industry to monitor the global network's state and identify potential malicious behavior. The industrial world has acquired much more experience in diagnosis and monitoring mechanisms that can inspire the security of intelligent transportation (autonomous cars); yet, the latter require their own particular needs and constraints related to reliability and their own system architecture.

The current evolution of transportation has been driven by innovative developments in numerous industrial and research fields, including machine learning (ML). These technical differences enable the acquisition and implementation of highly challenging tasks such as autonomous driving. This work aims at developing an intelligent anomaly detection system using ML in an automotive context. We intend to evaluate the potential of supervised learning in improving the safety of future autonomous vehicles. Our main objective is the reliability of the intersystem connections in the vehicle.

Specifically, this chapter focuses on detecting whether a sensor is anomalous (i.e., behaving abnormally) and accurately classifying the type of anomaly. We target anomalies that are caused by speed or sensor measurements taken, such as trajectories. The automated driving system can then quickly take mitigating actions once an anomaly is detected. To successfully identify and classify anomalies, we provide an automated approach based on ML. We exploit historical measurements collected from healthy and abnormal sensors to learn profiles of selected anomalies.

We compare the performance of different classification algorithms for evaluative purposes. We consider the following methods: logistic regression (LR), K-nearest neighbors (KNN), support vector classifier (SVC), naïve Bayes (NB), decision trees (DTree), and random forest (RF). In practice, the latter is the highest-scoring method for our data. Specifically, the contributions of this chapter are as follows:

- We introduce a new framework: an ensemble learning technique that uses different supervised classifiers to identify anomalies occurring in the vehicle interior system, based on readily available system monitoring data. Our model can detect anomalies with high accuracy.
- We demonstrate the effectiveness of our framework using a simulation system that involves injecting simulated anomalies into healthy sensor measurements. We report an F-score higher than 95% with this dataset.

The remainder of the chapter begins with an overview of related work (Section 2). The baseline methods used to develop the proposed framework are described in Section 3. Then, our anomaly detection and diagnosis framework is introduced in Section 4. Section 5 reports the experimental results and a subsequent detailed discussion. Section 6 ends the chapter with the main conclusions.

2 Related work

This section reviews related work in the area of vehicle system anomaly detection using ML techniques.

Anomaly detection and diagnosis in automated driving systems have been discussed in various publications. A predictive diagnosis based on supervised classification with labeled anomaly types included in the dataset is introduced by Jung et al. [2]. Such a model involves a representative learning set, i.e., it consists of combining model-based and data-driven approaches for fault diagnosis.

Al-Saud et al. [3] have proposed an intelligent method to empower electric vehicles with reliable anomaly detection and mitigation system. They have used an SVM algorithm based on a one-class identification process to monitor vehicle behavior. Moreover, they have adopted an optimization technique referred to as social spider optimization to adjust the classifier hyperparameters. Similarly, a comparative study of different ML techniques for anomaly detection in a smart vehicle is provided by Al-Saud et al. [4]. They have implemented and analyzed ML algorithms that allow anomaly recognition in various existing vehicle architectures, along with considering new designs in terms of plug-in module and software's update.

A detection model based on ML algorithms has been introduced by Park and Choi [5], where they have attempted to reduce the time of analysis along with maintaining the detection accuracy. They have implemented the model to detect adware and malware in the automated driving system by defining the architecture of intrusion detection. They have considered the prediction accuracy and the cost of each algorithm adopted to assess the effectiveness of their framework in preventing the malfunctioning of the vehicle.

By employing a set of measurements from vehicle sensors, an anomaly detection method has been introduced to identify abnormal behaviors of a vehicle to prevent potential faults [6]. The authors of this work have built a detector that consists of classifiers of one-class and two-class who learn from a training set and predict the occurrence of both previously seen and new types of anomaly.

A model for anomalous behaviors detection has been developed by Park and Choi [7]. The authors have applied a hierarchical data analysis for attack data classification. They have proposed a multilabeled hierarchical classifier to improve the accuracy of the detection of anomalies in the in-vehicle network. They designed a learning model that adapts to multiclass labeled schemes to simultaneously incorporate multiple attributes while classifying different types of attack data. In evaluating the performance of their model, they applied four ML algorithms to previously published models and contrasted various evaluation metrics, including accuracy, precision, recall, and the time required for the training and testing phases.

Lushan et al. [8] have considered unusual behavior and unexpected pattern of driving hardware systems as an anomaly to develop detection techniques that learn from input data, including GPS records, velocity, and the weight of the vehicle.

Recently, Francia and El-Sheikh [9] have illustrated the applicability of ML policies to the identification of anomalous network packets and vehicle patterns. They showed that vehicle pattern identification could lead to a better understanding of signatures of internal network traffic, which are frequently owned and not accessible to the public, and exploring various ML technologies can open the possibility of advancing artificial intelligence science in the automated driving field.

This work focuses on the employment of supervised ensemble-learning techniques for anomaly detection and classification. The notions of classification can have different meanings. In this chapter, we are interested in data-driven classifiers for binary classification, in which we verify whether the sensor is anomalous or healthy, along with a multiclass classification to identify the type of anomalous behavior that a sensor presents.

3 Baseline methods

In this section, we provide brief description of the baseline supervised algorithms used to build the proposed model.

3.1 Logistic regression

LR is a traditional and well-established method of supervised classification [10]. It is generally regarded as an extended version of ordinary regression and can be limited to modeling a dichotomous variable typically expressing the presence or absence of an event. An LR model predicts the probability of an event occurring (value of 1) or not (value of 0) based on the optimization of the regression coefficients. The output continuously varies between 0 and 1. When the predicted value is above a threshold, the event is likely to occur, while when the value is below the same threshold, it is not. The whole problem of classification by LR is a simple optimization problem where one tries to obtain the best set of parameters for a curve (often a sigmoid function) to fit the data. This is where the ML concept comes in.

3.2 Naïve Bayes

NB is one of the most widely used algorithms for classification purposes [11]. It refers to a simple type of classification based on Bayes' theorem, which is itself a basic result of probability theory. The mathematical principle consists of strong independence of assumptions [12]. These naive hypotheses imply the independence of the variables. The basis for this assumption is the necessity of generating multivariate probability values based on the training data. Practically, not all combinations of attribute values occur in the training data, nor are they sufficiently available. Therefore, estimating each relevant multivariate probability directly may not be entirely accurate. The conditional independence assumption allows NB to circumvent this limitation. The NB, regardless of the severe independence constraint, represents a powerful alternative classification tool in various real-world contexts [13].

3.3 Support vector classifier

SVC is a type of learning algorithm initially defined for discrimination, that is, the prediction of a binary qualitative variable. It was then generalized to the prediction of a quantitative variable [14]. In the case of discrimination of a dichotomous variable, it is based on the search for the optimal margin hyperplane, which, when possible, correctly classifies or separates the data while being as far as possible from all the observations. The aim is to find a classifier, or a discrimination function, whose generalization capacity (predictive quality) is as high as possible. In mathematical terms, the SVC finds the hyperplane whose distance to all examples is the highest possible. The hyperplane must be equidistant from the examples of the two most closely adjacent classes. The twofold distance between the hyperplane and the points closest to it is the margin in SVC terminology. Accordingly, SVC is also referred to as a margin-maximizing classifier. Margin maximization, or selecting a particular equivalent hyperplane between examples of two separate classes, is critically important. This provides a convenient generalization mechanism for newly seen examples classification.

3.4 K-nearest neighbor

KNN is a standard ML method that has been extensively applied to large-scale data mining tasks [15]. It is based on the idea that from a labeled dataset, one can estimate the class of a new input by examining the majority class of the KNN (hence the name of the algorithm). The only parameter to set is K, the number of neighbors to consider. The most commonly chosen metrics are the usual Euclidean distance and the Mahalanobis distance (which takes into account the variance – from a statistical point of view – and the correlation between the data). Although the algorithm can work with these default metrics, it is generally much better when a metric adapted to the data is used, a metric that can be computed from known heuristics related to the problem (e.g., the weighted Euclidean distance) [16].

3.5 Decision trees

In the ML community, DTrees are recognized as a potential tool for classification purposes. They become increasingly widespread due to their ability to tackle complex tasks by providing a comprehensible representation that is easier to interpret, as well as their suitability for the task of inference by providing logic-based rules for classification [17]. This decision or data mining tool allows representing a set of alternatives in the form of a graphical tree. Concretely, DTrees models a hierarchy of tests to predict a result. The possible decisions are found at the edges of the branches (the "leaves" of the tree) and are reached according to decisions made at each step. A DTree works by iteratively applying logical rules (typically data separations by "hyperplane," generalization of a plane with more than two dimensions), each rule being chosen according to the result of the previous rule. The advantages of DTrees include simplicity of interpretation, speed of training, nonparametricity, and the fact that they require very little preprocessing of the data. They can be automatically computed by supervised learning algorithms capable of automatically selecting discriminating variables within unstructured and potentially large data. These algorithms can also be used to extract logical rules that do not appear in the raw data. Another use in ML consists of building not a tree but a forest of DTrees. A decision is then made by "voting" on all the trees and choosing the dominant answer (for a discrete choice) or the average of the answers (for a continuous variable). The results obtained in this way are significant, especially when the DTree are used in RFs [18].

3.6 Random forests

The RF algorithm was proposed by L. Breiman and A. Cutler in 2001 [18]. In its most classical form, it performs parallel learning on multiple randomly constructed DTrees trained on different subsets of data. The ideal number of trees, which can go up to several hundred or more, is an important parameter: it is highly scalable and depends on the problem to be solved. Concretely, each tree of the RF is trained on a random subset of data according to the bagging principle, with a random subset of features (variable features of the data) according to the "random projections" principle. The predictions are then averaged when the data are quantitative or used for a vote for qualitative data, in the case of classification trees. The RF algorithm is known to be one of the most efficient "out-of-the-box" classifiers (i.e., requiring little data preprocessing).

4 Proposed framework

Figure 1 provides an overview of the proposed framework. It is a mechanism for identifying anomalies in driving systems from the monitoring data collected from the vehicle. The monitoring inputs are captured from individual sensors (the in-vehicle speed sensor, GPS speed sensor, and in-vehicle acceleration sensor), which the density distribution is exhibited in Figure 2. Next, we produce statistical features to reduce the amount of computational and storage costs while preserving relevant information in the measurements. Ultimately, we train a classifier for every single sensor anomaly to distinguish it from the overall anomalies using labeled data history. The proposed framework can analyze the monitoring data and labels the anomaly of each sensor based on the classifiers' predictions.

4.1 Data acquisition and monitoring

We utilize data already captured periodically in CAV transportation to assess the overall system health. Typically, these data comprise (CAV) trajectories, including anomalies in sensor measurements. Our framework does not depend on a specific set of collected measurements and can be paired with a variety of smart driver models.

Because of the lack of a publicly available CAV trajectory dataset incorporating anomalies in the sensor measurements, the dataset for this study is produced using a simulation framework and constructed by adding random anomalies to the normal trajectory data. The four main types of injected signals, including short circuit, noise, bias, and gradual drift, are included, resulting in four main types of anomalies denoted, in this work, as bias, drift, constant, and instant anomalies, which are described as follows:

- *Instant*: A sudden and unexplained change in the monitored data occurring within two successive sensor readings, which can be considered as an instant anomaly. This kind of anomaly pattern is modeled by a Gaussian random variable with a mean of 0 and a given variance.
- *Constant*: It refers to an accumulation of variance in the sensor temporarily constant readings, which does not correlate with the physical phenomena involved. This type of malfunction, unlike short, is produced on multiple successive sensor recordings in a way that it may be assumed to be constant. It is modeled as a series of independent and identically distributed Gaussian random variables.

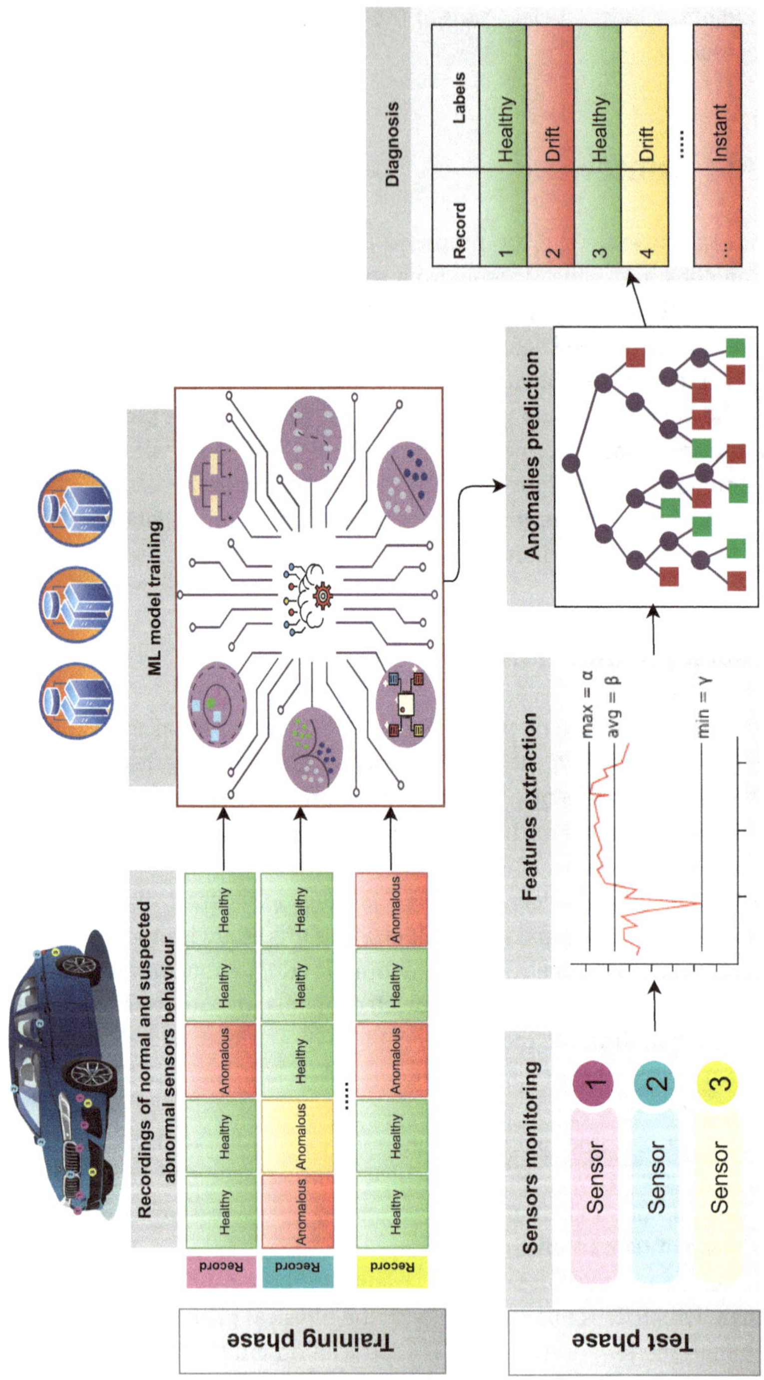

Figure 1: An overview of the proposed ML model.

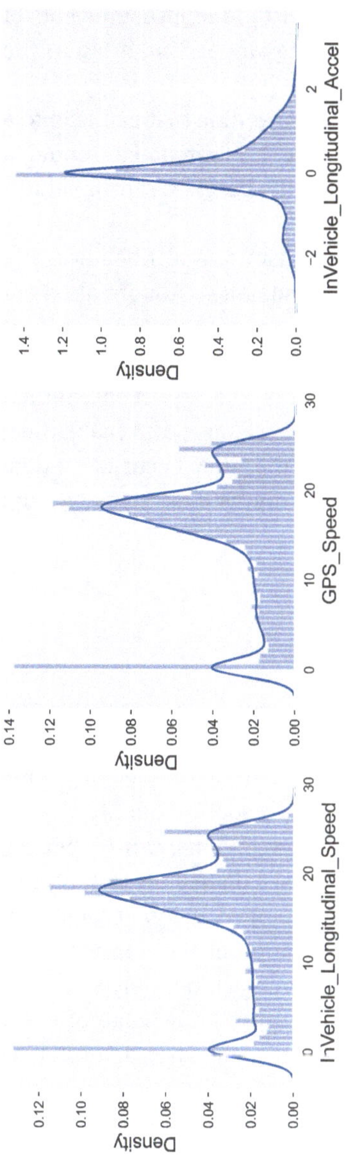

Figure 2: The density distribution of the most relevant features captured by the vehicle sensors.

- *Bias*: A shift that is temporarily maintained with regard to the measurements of the sensor. This kind of anomaly bias is modeled by appending a lag to the observation.
- *Drift*: It refers to a slight shift in the observable data over time that can lead to a significant deviation in the observed system over time. It is simulated through a set of linearly increasing/decreasing values added to the baseline sensor values.

Following the simulation model proposed by Wang, Masoud, and Khojandi [19], each sensor statement is inflected by the four types of anomalies. Abnormal sensor occurrences are assumed to happen independently. This means that the classifiers are not explicitly trained using a dataset with either cross-linked sensor malfunctions or system-level attacks on car detectors. To modify and adjust the normal measurements of the corresponding sensors in the initial dataset and make them anomalous, abnormalities are simulated in such a way that they occur at random times, during random periods of time (if applicable for the type of anomaly), and are assigned to randomly selected sensors.

4.2 Classification of the synthetic anomalies

The first purpose of the framework is to learn and recognize the symptoms of anomalies that have been previously detected. To assess our proposed framework, designed synthetic anomalies emulating typical observed behavioral defects in the automated driving system are classified. We train an ML model to identify a given set of anomalies through a training set of these labeled exceptions. Our framework labels each instance with the respective anomaly or can also label new instances as healthy when no anomaly is recognized. We use a one-to-one version of the classifier for each classification: that is, for each implementation of the training set, a distinct classifier is formed to identify the defect. This approach not only simplifies adding new anomalies into the classifier set and acquiring knowledge of every anomaly's nature, but it also provides a way to train only for the anomalous behaviors we are concerned with.

The second purpose is to identify faults that lead to fluctuations in system operation, the triggering of a false fault alarm can lead to disruptions in system operation or even crashes if the fault alarm triggers an automated corrective action. Therefore, it is always important to prevent false alarms rather than miss abnormalities. To improve scalability and reduce false alarms, a sensor is not predicted to be anomalous using the data obtained from a single trajectory. We, instead, we assume an anomaly prediction to be successful when the same prediction occurs for C consecutive trajectories. Otherwise, we classify the state as healthy. C varies according to the target anomaly features along with the host system. Accordingly, C is determined by trial and error on preliminary experiments.

5 Experimental results and discussion

In this section, we first present the implementation details of this model. Then, we compare the anomaly detection algorithms, adopted for the automated driving system diagnosis, in terms of accuracy and robustness in both training and test phases. Also, we discuss the impact of the confidence threshold introduced previously (C). Finally, we investigate the metric of detection delay involved for each supervised algorithm used.

5.1 Implementation details

We implement the proposed framework in python. We use the *Scikit-learn* package for the six supervised learning algorithms. In *Scikit-learn*, the default hyperparameters of such classifiers are adjusted for a smaller dataset than ours. Hence, we first apply a systematic grid search to find the best values of the hyperparameters for the algorithms considered in the classification task. These hyperparameters can be the number of trees for tree-based approaches (DTree and RF), the number of neighbors and the distance used for KNN, the kernel involved for SVC and several other hyperparameters. We use the *GridSearchCV* introduced in the *Scikit-learn* package that consists of comparing the resulting models with all combinations of values for the hyperparameters.

5.2 Evaluation

We evaluate the ability of our framework to identify anomalies with different velocity fluctuations. First, we examine the classification performance in identifying known anomalies with new input configurations. Second, we evaluate the system's labeling ability for incoming unknown anomalies.

We initially undertake a fivefold cross-validation for all tests,, in which we divide the dataset into five sets that have equal distributions of anomalies within the original dataset. We next train five separate instances with four sets. In testing, we run the fifth set that has been retired from the training set.

For the outputs, we report a commonly utilized metric of classification performance, the F-score. Specifically for binary classification, this metric is determined by taking the harmonic mean of accuracy and recall, where the accuracy is given by the proportion of true positives to the number of total positive predictions, and the recall is given by the proportion of true positives to the number of total true positives in the dataset. The F-score ranges from 0 (worst) to 1 (best). Since all the results obtained are multiclass, we compute the mean accuracy and recall for each individual class and consider the harmonic mean to compute the aggregate F-score.

5.3 Confidence threshold factor

To minimize the false anomaly warnings, all nonpersistent misclassifications are eliminated by imposing a confidence threshold, C. We consider the prediction as effective only if it remains persistent throughout C consecutive records. The evolution of the anomaly false alarm rate (Figure 3a) and the misidentification rate (Figure 3b), depending on different values of confidence threshold C, are illustrated in Figure 3.

We observe that the false alarm rate begins decreasing from $C = 15$ for most of the considered classifiers. For example, the false alarm rate caused by DTree is reduced from 20% to 6%. On the other hand, $C = 15$ can be considered as the stabilization point after a significant increase of the anomaly miss rate. Thus, a confidence threshold of $C = 15$ is applied for the remainder of the chapter to ensure a low delay in anomaly identification along with a low false alarm rate for all classifiers.

5.4 Anomaly detection results

Figure 4 displays the F-scores for the anomaly types we investigate for the six classifiers used in the training and testing phases (Table 1). From the analysis of the figure, it follows that the accuracy of the majority of the algorithms considered gets lowered, especially for the KNN and RF classifiers. Moreover, the F-scores that correspond to the *drift* tend to be lower than the scores for other anomalies because the classifiers are prone to incorrectly predict the drift as healthy (and vice versa). This can be explained by the fact that its behavior is similar in some trajectories. Owing to the random selection of features in the RF approach, it also employs the features that do not serve as the major anomaly predictors, yet they are still beneficial for anomaly recognition. As a result, the RF performs better against noise in the data, making it consistently effective at detecting drift where other classifiers may fail.

On the other hand, Figure 5 illustrates the fake positive and negative rates for abnormalities recognition. In terms of missed anomaly rate, RF and LR systematically outperform the other techniques and classify defects almost perfectly (less than 5%). However, the worst overall performance is obtained with SCV, which misses more than 25% of abnormal recordings and misclassifies the target anomalies. It generates false anomaly alerts for more than 6%, both of which can be lowered by adjusting the C-parameter which can minimize the missed anomaly occurrence and the faulty alerts. All the same for NB algorithm that misses nearly 20% of the anomalies. This can be due to the assumption of independent predictors that NB approach adopts. In fact, it implicitly assumes that every attribute is mutually independent. However, in practice, it is nearly unattainable that all predictors are completely independent from one another.

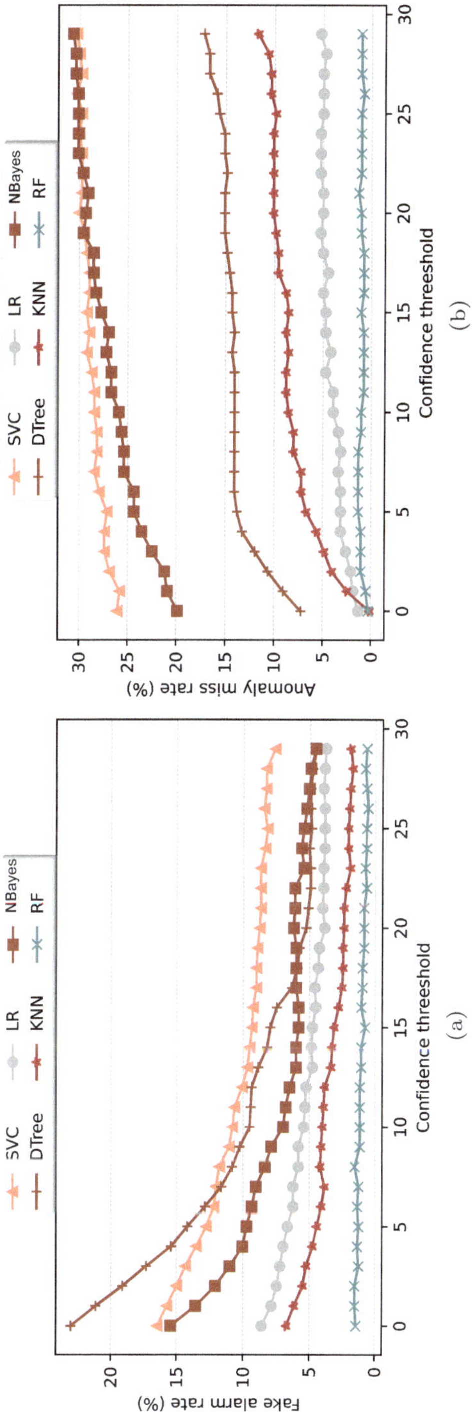

Figure 3: The variation of (a) the fake alarm rate and (b) the anomaly miss rate depending on the variation of the confidence threshold C.

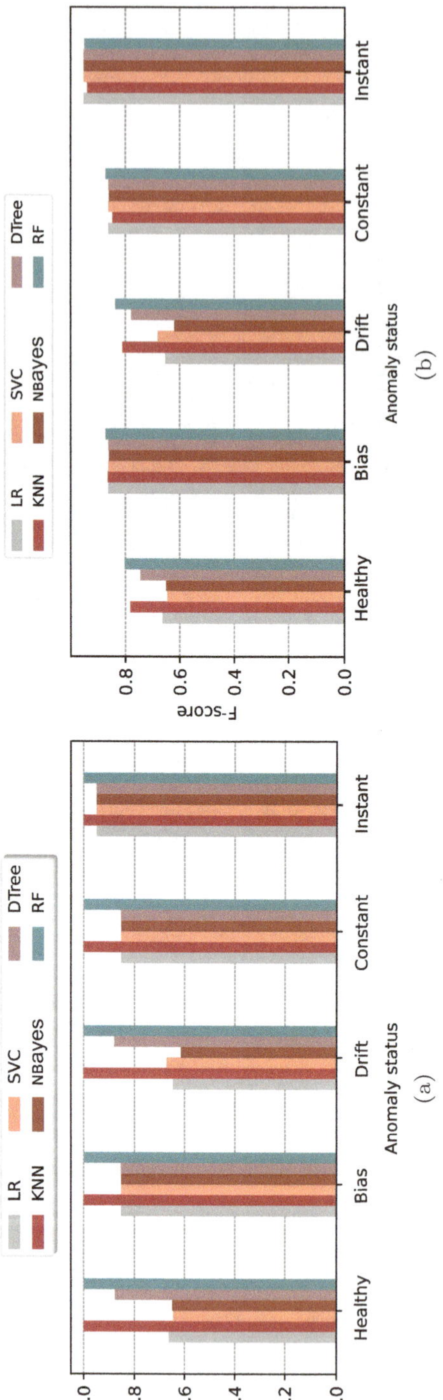

Figure 4: The F-score statistics reached by the classifiers used during (a) the training phase and (b) the test phase.

Table 1: The F-score statistics reached by the classifiers used during (a) the training phase and (b) the test phase.

		LR	KNN	SVC	Naive Bayes	DTree	RF
Training	**Healthy**	0.661	0.999	0.645	0.648	0.876	0.999
	Bias	0.852	0.999	0.852	0.852	0.853	0.998
	Drift	0.646	0.999	0.671	0.614	0.878	0.999
	Constant	0.852	0.999	0.852	0.852	0.853	0.998
	Instant	0.948	0.999	0.948	0.948	0.948	0.999
Test	**Healthy**	0.664	0.78	0.646	0.651	0.745	0.803
	Bias	0.861	0.863	0.861	0.859	0.861	0.871
	Drift	0.653	0.810	0.68	0.619	0.779	0.837
	Constant	0.861	0.846	0.861	0.859	0.861	0.871
	Instant	0.95	0.938	0.95	0.951	0.95	0.949

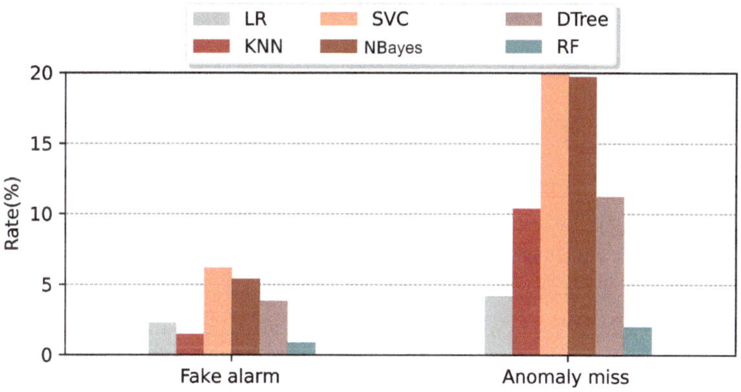

Figure 5: The rate of positive and negative anomaly recognition obtained by the six classifiers.

More interestingly, the most significant features are defined as ones that successfully separate different classes in the training data, and the *Scikit-learn* Python package denote them through the normalized *Gini reduction* afforded by each one. The most important features in our models are computed based on sensor measurements such as GPS speed and vehicle acceleration, which have been shown to be relevant to the types of anomalies considered in this work.

5.5 Anomaly detection delay

To study the detection delay, a special random shift of anomalies is used. The anomaly begins at a time chosen at random during the driving process. In this work, we simulate this operation by injecting anomalies along different consecutive recording windows. For each classifier, we capture the time delay between the time when the anomaly starts and the first window, where sequential windows get tagged as anomalous with the appropriate type of anomaly.

The observed detection delay for each classifier and anomaly (Figure 6a), together with the percentage of anomalies that are not recognized before the end of the simulation (Figure 6b), is exhibited in Figure 6. The RF algorithm provides no more than a 5–10 s delay when diagnosing *drift* and *instant* anomalies. For other anomalies, there is a higher delay, wherein the analysis consists of statistics that develop gradually as the 15 s measuring window shifts. SVC involves a significant delay to detect anomalies, which may be explained by the noise occurring in the dataset used. Curiously enough, a noticeable percentage of undetected *instant* anomaly is reported, which may be due to the nature of the anomaly that refers to the variation of the monitored data between two consecutive sensor readings.

6 Conclusion

Automatic detection and diagnosis of anomalies in autonomous vehicles have become increasingly prominent in today's rapidly expanding transportation systems. In this chapter, we proposed an anomaly detection and diagnosis framework, which leverages the already monitored data collected from four sensors in a vehicle to learn and identify the profiles of previously observed anomalies.

We evaluated the proposed framework using simulations consisting of injecting different anomaly signals to healthy records gathered using velocity and acceleration sensors. We demonstrated that our model accurately identifies 98% of the synthetic anomalies. The percentage of false anomaly alarms differs from one classifier to another and following the type of anomaly monitored.

In this chapter, we have focused on detecting different profiles of synthetic anomalies. In future work, we intend to focus on designing synthetic anomalies that closely mimic the anomalies experienced by a driving system in a vehicle based on other internal and external sensors. Furthermore, we will adopt more advanced ML algorithms such as neuroclassifiers and reinforcement learning techniques.

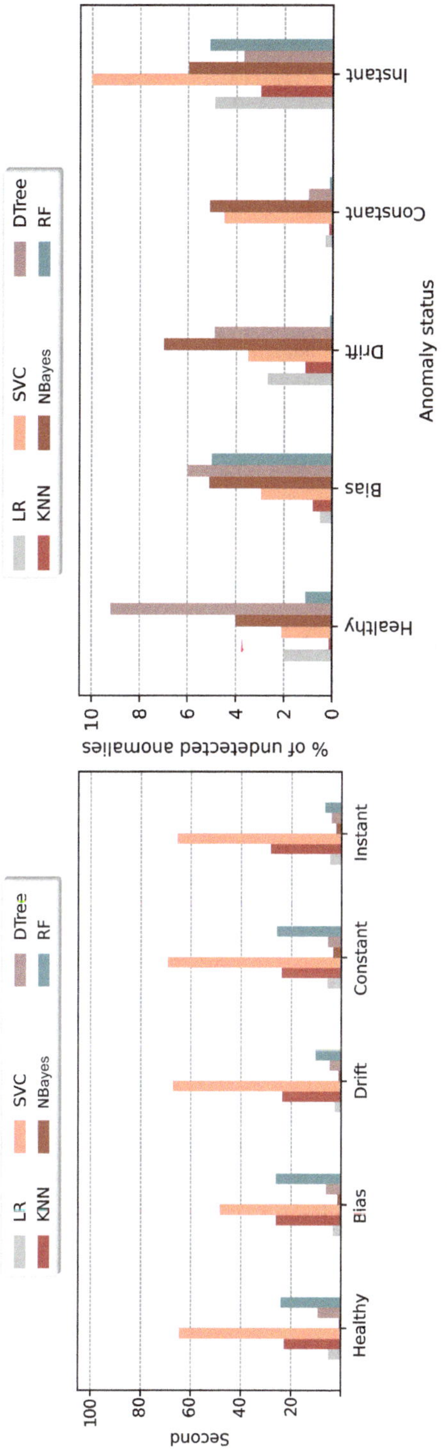

(a) Anomaly detection delay

(b) Undetected anomalies rate

Figure 6: The anomaly detection delay along with the rate of missed anomalies reached using the proposed model.

References

[1] Calvert SC, Mecacci G, van Arem B, de Sio FS, Heikoop DD, Hagenzieker M. Gaps in the control of automated vehicles on roads. In *IEEE Intelligent Transportation Systems Magazine*, 1–1, 2020.

[2] Jung D, Ng KY, Frisk E, Krysander M. A combined diagnosis system design using model-based and data-driven methods. In *2016 3rd Conference on Control and Fault-Tolerant Systems (SysTol)*, 177–182, 2016.

[3] Al-Saud M, Eltamaly AM, Mohamed MA, Kavousi-Fard A. An intelligent data-driven model to secure intravehicle communications based on machine learning. *IEEE Transactions on Industrial Electronics*, 67 (6):51125119, 2020.

[4] Berger I, Rieke R, Kolomeets M, Chechulin A, Kotenko I. Comparative study of machine learning methods for in-vehicle intrusion detection. In Katsikas SK, Cuppens F, Cuppens N, Lambrinoudakis C, Antón A, Gritzalis S, Mylopoulos J, Kalloniatis C editors, *Computer Security*. Springer International Publishing, 85–101, 2019.

[5] Park S, Choi J-Y. Malware detection in self-driving vehicles using machine learning algorithms. *Journal of Advanced Transportation*, 2020:3035741, 2020.

[6] Theissler A. Detecting known and unknown faults in automotive systems using ensemble-based anomaly detection. *Knowledge-Based Systems*, 123:163–173, 2017.

[7] Park S, Choi J-Y. Hierarchical anomaly detection model for in-vehicle networks using machine learning algorithms. *Sensors*, 20(14): 2020.

[8] Lushan M, Bhattacharjee M, Ahmed T, Rahman MA, Ahmed S. Supervising vehicle using pattern recognition: Detecting unusual behavior using machine learning algorithms. In *2018 IEEE Region Ten Symposium (Tensymp)*, 277–281, 2018.

[9] Francia GA, El-Sheikh E. *Applied Machine Learning to Vehicle Security*, 423–442. Springer International Publishing, Cham, 2021.

[10] Hosmer DW Jr, Lemeshow S, Rodney X. Sturdivant. In *Applied Logistic Regression, 3rd Edition*. Wiley, 2013.

[11] Wu X, Kumar V, Ross Quinlan J, Ghosh J, Yang Q, Motoda H, McLachlan GJ, Ng A, Liu B, Yu Philip S, et al. Top 10 algorithms in data mining. *Knowledge and Information Systems*, 14(1):1–37, 2008.

[12] Duda RO, Hart PE, Stork DG. *Pattern Classification and Scene Analysis*. volume 3. Wiley, New York, 1973.

[13] Zhang L, Jiang L, Chaoqun L, Kong G. Two feature weighting approaches for naive bayes text classifiers. *Knowledge-Based Systems*, 100:137–144, 2016.

[14] Hearst MA, Dumais ST, Osuna E, Platt J, Scholkopf B. Support vector machines. *IEEE Intelligent Systems and Their Applications*, 13(4):18–28, 1998.

[15] Nadkarni P. Chapter 10 – core technologies: Data mining and “big data”. In Nadkarni P editor, *Clinical Research Computing*. Academic Press, 187–204, 2016.

[16] Alfeilat HAA, Hassanat ABA, Lasassmeh O, Tarawneh AS, Alhasanat MB, Eyal Salman HS, Prasath VBS. Effects of distance measure choice on k-nearest neighbor classifier performance: A review. *Big Data*, 7(4):221–248, 2019.

[17] Qualitative classification with possibilistic decision trees. In Bouchon-Meunier B, Coletti G, Yager RR editors, *Modern Information Processing*. Elsevier Science, Amsterdam, 159–169, 2006.

[18] Breiman L. Random forests. *Machine Learning*, 45(1):5–32, 2001.

[19] Wang Y, Masoud N, Khojandi A. Real-time sensor anomaly detection and recovery in connected automated vehicle sensors. *IEEE Transactions on Intelligent Transportation Systems*, 22(3):1411–1421, 2021.

Mohan Krishna Kagita, Mohammed Kaosar, Navod Thilakarathne

Survey on AI-based IoT and drone-equipped smart agriculture

Abstract: Over the past few decades, there are rapid advancements in technologies where massive intelligence applications came into being. These developments assist in the development of intelligent capabilities that prompt these systems to communicate among themselves. Artificial intelligence (AI) and Internet of things (IoT) remain some of the widely evolving modern intelligent concepts. These technologies connect multiple things across the Internet triggering them to make decisions like humans. The application of AI and IoT continues to evolve where specific industries boast their potential implications. The article discusses how AI and IoT continue to revolutionize agriculture with the application of smart agriculture. AI and IoT's involvement in agriculture makes it easy to make predictions like temperature, rainfall, humidity, soil statuses among others. Agriculture's AI and IoT continue to advance farming practices by bringing profitable ventures in the end.

Keywords: artificial intelligence, Internet of things, smart agriculture, farming agriculture

1 Introduction

Agriculture is experiencing a new technological transition, which looks optimistic, as it enables it to go into the next level of production and profitability [1]. Smart farming, which is the use of input variables when and where necessary, is the new wave of the modern farming revolution (one mechanized farming, the second agricultural revolution with genetically modified and now, by the fact that more data is available) and it is being reinforced by increasing the farm knowledge systems. In October 2016, the United States Department of Agriculture (USDA) claimed that innovative agriculture technologies improved portfolio income and operating earnings [2]. New technologies are also being implemented in ranches to ensure the growth of agricultural production in the context of environmental considerations.

Mohan Krishna Kagita, School of Computing and Mathematics, Charles Sturt University, Melbourne, Australia, e-mail: m.kagita@cqumail.com
Mohammed Kaosar, Discipline of Information Technology, Media and Communication, Murdoch University, Australia, e-mail: mohammed.kaosar@murdoch.edu.au
Navod Thilakarathne, Department of ICT, University of Colombo, Sri Lanka, e-mail: navod.neranjan@ict.cmb.ac.lk

https://doi.org/10.1515/9783110760477-007

The use of these technologies, however, implies unpredictability and market uncertainties as artificial intelligence (AI) is constantly evolving. The current agricultural developments have brought AI-based machinery and equipment to various levels. This knowledge has increased crop output and enhanced monitoring, harvest, handing out and packaging in real time. The latest automatic identification technologies using farming robots and drones have enormously impacted the agro-based industry. Different computer-based high-tech systems are created to determine critical factors such as weed discovery, yield detection, crop eminence, and numerous other approaches [3].

As per a marketing plan, higher education and training of agriculturalists, shared information, accessibility of fiscal incomes, and increased consumer preferences for natural food are elements that could support sustainable agriculture [4]. Collecting data from crops using these new technologies is to develop something cohesive and worthwhile as data alone is not applicable, only numbers or photos. Farms that elect to be technologically driven have tremendous benefits, such as saving money and work, increasing productivity, cutting costs with minimum effort, and providing excellent food with ecologically friendly techniques. However, the use for the farm will rely not only on farmers' desire to implement new technology in their sector but also on each agricultural potential in terms of economies of scale with a profit margin that increases with the farming size. The USDA reported that, on average, smart agriculture adopters had an annual margin of 170 dollars per unit of land higher than those not adopted, bearing in mind that for three techniques, the highest adoption levels were over 1,700 hectares for ranches [2]. Depending on the crop, these margins may reach up to $300. Increased usage of intelligent agricultural services is essential to improve financial performance and address a growing population's food needs.

2 AI's impact on agriculture

AI-based technologies increase efficiency in all sectors and address the issues of many businesses, as well as the different fields of farming such as crop yield, irrigation, the composition of the soil, crop management, weeding, and plant establishment [5]. To supply highly valued applications of AI in the industry, agricultural automation is recommended. Agriculture is experiencing a crisis in the world, but AI can attain better results. AI-based technical solutions allow farmers to increase production with fewer investments, improve output quality, and make the returned harvests more accessible to the market. By 2021, 80 million linked gadgets will be used by farmers. By 2050, the average farm will create an average of 5 million pieces of information per day.

In an agricultural environment, the Internet of things (IoT) involves various sensors to turn all elements and actions engaged in agriculture into information. Estimates of about 10% to 15% of US farmers on the farm use IoT solutions for 1,400 million hectares, and 200,000 farms were mentioned [6]. It is projected that the agricultural industry will be heavily influenced by the progress made on these technologies [7]. It is approximated that IoT is likely to increase its agriculture production by 70% by 2050 with new technologies [8], and that is positive even though, according to Albaji et al. [9], by 2050, the agriculture needs to raise by 55% due to an increasing population over 10,000 million. The key benefits of IoT use are increased yields and lower costs. For instance, from Figure 1, in On-Farm research, the output on an average farm utilizing IoT increases by 2% and energy expenditures are reduced between 15 and 30 dollars per hectare. In contrast, irrigation uses are reduced by 8%. The prevalence of IoT devices for agriculture is shown in the following table.

The following are the several ways in which AI contributed to the agriculture sector:

2.1 Competencies and labor force

According to Giles et al. [10], AI enables farmers to collect large amounts of data from policy and social websites, evaluate it, and provide farmers with alternatives to numerous ambiguous issues. It also allows us to use irrigation more intelligently, resulting in increased yield for agricultural producers [11]. Due to AI, agriculture, in the coming years, will be discovered to be a combination of technology and physical skills, which will result in a higher quality product for all farmers and reduce their expenses and burden. According to the UN, by 2050, two-thirds of the world's population will live in cities, necessitating the need to ease the load on farmers. AI may be used in agriculture to automate numerous operations, mitigate risk, and give farmers a straightforward and effective farming system.

2.2 Optimize the output

In their study, Ferguson et al. [12] mentioned that selection and the quality of the seed set the highest degree of performance for all plants. The new technologies have enabled us to choose crops best and have even increased the development of hybrid seeds that are best adapted to farmers' demands. It understands how the seeds adapt to various seasonal changes and different types of soil. The odds of plant illnesses are lowered by gathering this information. Now, we are prepared to fulfil market trends, annual results, and consumer needs so that farmers can optimize crop returns economically.

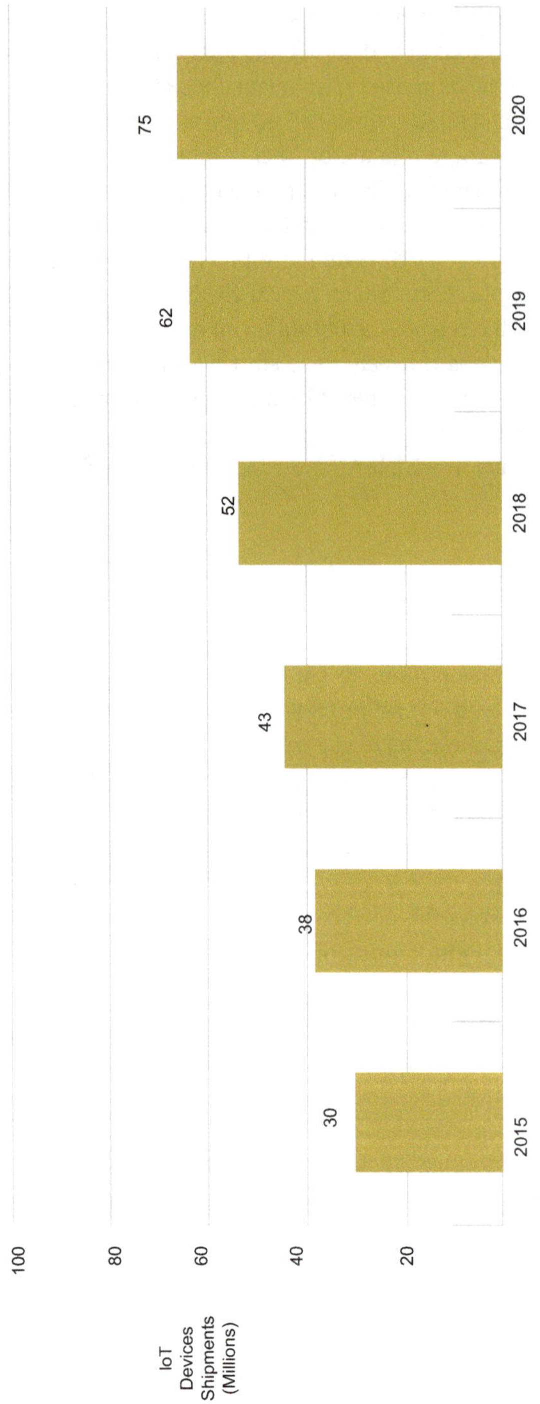

Figure 1: Estimate of IoT shipments global.

2.3 Farmers' chatbots

Chatbots are interactive virtual assistants that automate user engagement. Chatbots AI-based, combined with machine learning approaches, have allowed us to comprehend natural language and communicate with people more personalized manner. They are primarily utilized for retail, travel, and media. Still, agriculture has benefited from this facility by supporting farmers in obtaining answers to unanswered questions, providing guidance, and making various proposals [13].

2.4 Recognize and comprehend images

According to Lee et al. [14], there has been an increase in interest in autonomous uncrewed aerial vehicles (UAVs) and their applications, including recognition and monitoring, human body identification and positioning, rescue operations, and bushfire detection [15]. Due to its wide range and incredible imaging technology, which ranges from delivery to photography, the capacity to be controlled remotely via remote controller, and the systems' agility in the air, which facilitates us to accomplish a great deal with this equipment, drones or UAVs are becoming increasingly popular for reaching great heights and intervals and performing a variety of tasks.

3 Agriculture robots

In broad areas of economics with remarkably low efficiency, such as farming, robotics and autonomous systems are implemented. According to Ducktt et al. [11], the UK Agri-Food chain delivers more than £110 billion annually from primary agriculture to retail, while 3.5 m in a genuinely international business yields £25 billion in exports in 2016. Robotics have played an essential part in the invention and managing of agricultural products. Recently, researchers have begun emphasizing technology for developing autonomous agricultural tools because conservative farming equipment needed effectiveness [16]. The primary aim of this technology is to substitute human labor and generate practical benefits for both small and large productions. In this field, robotic technology has dramatically enhanced productivity [17]. The robots undertake multiple agricultural activities autonomously, such as weaving, Irrigation, monitoring farms to produce effective updates, ensuring that adverse climatic conditions do not impact production, increasing precise accuracy also managing individual plants in different unusual conducts.

The concept of developing like a technology arose with a machine named cotton gin invented be Eli Whitney. In 1794, Eli Whitney (1765–1825), who was born in the USA, invented an instrument that transformed cotton manufacturing to remove seed from cotton fiber. In one day, it manufactured 50 pounds of cotton. Thus, autonomous agricultural robots were born. A simple automated model for determining the exact position of seeds was developed. Ultrahigh-precision seed planting has also been achieved. Mechanisms to ensure that sown seeds are zero ground speed [18]. This is vital because the seed does not bounce after the soil hit from its actual position. Automated machines documented the status or growth of the plant. Different biosensors were developed for plant growth monitoring and detecting plant diseases [19]. The traditional weaving technique was replaced by laser wearing technology, where portable infrared-light damage weed cells, computers drove the beam. Automated irrigation systems were also designed for the active usage of water.

The capacity of robotic farming machinery to improve yields has brought the agribusiness industry to increase. Tractica's analysis by market analysts and consultants indicated that global robotic shipments built for agricultural application will increase to $87.9 billion by 2025, as shown in Figure 2.

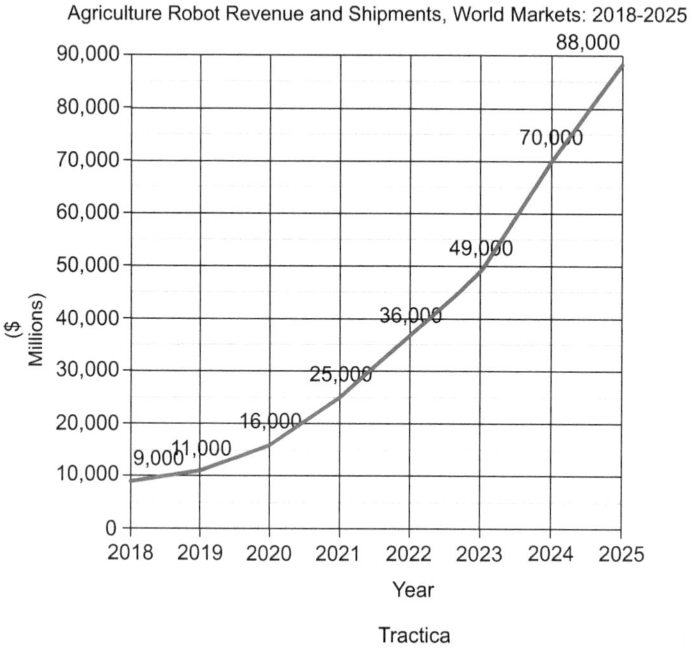

Figure 2: World market of agriculture robot revenue and shipments.

4 Robots used in irrigation

Agriculture absorbs 85% of the world's available freshwater resources. And this percentage is growing fast as the population grows and the need for food increases. This means that we need more efficient methods to ensure the proper use of irrigation water supplies. Automatic planning approaches replaced manual irrigation based on soil water measurements. During the operation of autonomous irrigation systems, the evaporation and transpiration of the plant depending on several atmosphere elements such as humidity, wind velocity, sunlight, and even crop variables like development stage, plant density, soil attributes, and pests was considered.

Analyses were done on various irrigation systems to build a system that uses fewer resources and increases efficiency [20]. On the farm, devices such as fertility meters and pH meters are used to determine soil fertility by measuring the proportion of main soil elements such as potassium, phosphorus, and nitrogen. Automatic plant irrigators are installed on the ground by wireless drip irrigation technology. This procedure ensures the productivity of the soil and ensures that water resources are used effectively.

Intelligent irrigation technology is developed to boost productivity without the participation of a large number of people, measuring water levels, soil temperature, nutrient content, and weather forecasts. The actuation is carried out by turning the irrigator pumps ON/OF as per the controller. The machine-to-machine technology is designed to facilitate communication and sharing of information between each other and among the host and the cloud via the central system between all the units of the agricultural industry [21]. They (2017) have built an automated robotic model to identify the Arduino and Raspberry pi3's humidity and temperature levels. The inputs are sensed at periodic intervals and supplied to the Arduino microcontroller (which has a hardware edge level). The signal is transmitted to the Raspberry pi3 (embedded by the K-nearest neighbor algorithm), and the signal is sent to Arduino to begin the irrigation system. The resource gives water as per the need, and the sensor readings are also updated and stored. Huang et al. [22] have also created an Arduino technology automated irrigation system for decreasing human power and time usage in the irrigation process. The irrigator pumps thus function on the information of the soil's qualities (Figure 3).

5 Agricultural drones

UAVs, otherwise known as automatons, are aircraft without a human pilot who may be automatically controlled [24]. They work in conjunction with the GPS and other sensors. Drones for crop monitoring systems, irrigation monitoring systems, weed detection, flock and animal monitoring, and calamity management are indeed

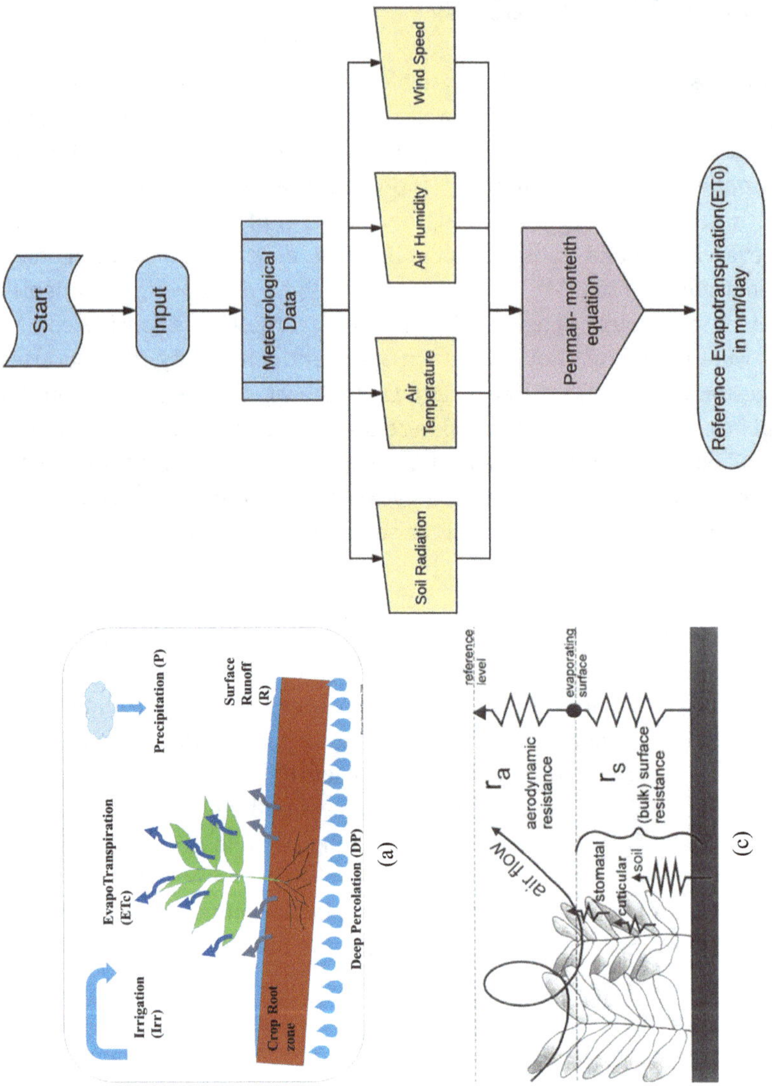

Figure 3: (a) Soil water balance evapotranspiration modeling components, (b) evapotranspiration flowchart [23], and (c) FAO Penman–Monteith method.

being deployed in agricultural production [25], as shown in Figure 4. Remote sensing, using UAVs to capture and analyze images, has a significant effect on agriculture. Rural businesses seem to have enthusiastically grasped robust innovation by using these driven instruments to improve conventional farming practices [26]. The total estimate of the automation systems in each vital sector is essential – over $130 billion, as the ongoing PwC investigation suggests. They can be opposed to a standard camera that is simple to use for unmixed images. Still, while a normal camera can give information about plant growth, incorporation, and various things, a hyperspectral sensor enables farmers to see items in a noticeable range, for example, ground moisture content and plant health surveillance. This could help to overcome the various constraints which hinder agricultural development [27].

The unmanned aerial systems (UAS) is developed with the wireless sensor networks (WSN). The data collected by WSN allows the UAS to promote its use, for example, by restricting synthetic compound spraying to carefully specified zones. Because the environmental factors are abruptly and continuously changing, the control cycle must almost likely respond as rapidly as anticipated. Conciliation with WSN can contribute to this journey [28]. UAVs are largely valuable for agricultural activities such as soil and field analysis, crop surveillance, crop height estimates, and pesticide spraying in smart agriculture [29]. Its hardware implementations adhere exclusively to critical criteria such as mass, flight range, payload, customization, and costs. Exploration of UAV technologies, methodologies, systems, and constraints is investigated [30]. About 300 concepts are studied and analyzed to identify a suitable UAV in farming [31]. In the following years, the agricultural drone business is predicted to increase by over 40%. It is thought that the necessity for efficient agriculture would only increase as populations levels increase and climatic patterns change [32].

The UAVs, sometimes referred to as drones, are primarily based on sensor and embedded system advancements designed to compensate for the pilot's absence and thus permit the movement of unmanned cars and their autonomous driving [33]. These drones have been employed by farmers as spray nozzles for many years and are regarded as effective and significant in foggy climates. They have also addressed the inaccessibility of a huge crop field, such as maize [34]. Additionally, they are recognized as having a distinct advantage over high-image satellite airborne detectors [35]. Giles et al. [10] modified an air carrier planting sprayer with a microprocessor-based sprayer control frame. A framework for the quantity of materials applied to the foliage was interfaced with a PC that controlled the three dose multipliers on either side of the sprayer via control calculations based on the amount of spray delivered. While IoT agricultural solutions are not so popular as consumer-oriented solutions, the market is fairly powerful. Numerous studies show that the interest in IoT in agriculture is steadily increasing.

Business Insider, for instance, predicts the shipment of 75 M agricultural IoT devices worldwide by 2020. And that sounds like a convincing statement that humankind must generate more food to feed the expanding population.

Figure 4: Agricultural drone types: (a) crop monitoring drone, (b) crop spraying drone, and (c) aerial mapping drone.

Updated a horticulture management circle by spraying synthetic materials on the products to which the drones were attached [36]. These drones were equipped with sensors that monitored how synthetic compounds were applied to crops in the field categorized as WSN. The data acquired by these remote sensors limited drones to spraying synthetic chemicals only in certain areas. Designed a low volume sprayer for an autonomous aircraft [37]. The helicopter employed in this study has a main rotor range of 3 m and a payload capacity of 23 kg. One barrel of gas was engaged for about 45 min. This strategy and the systematic results of this technique give an antecedent that might be used to create UAVs flying relevant areas with more excellent goal rates and larger droplet dimensions.

Advanced sensors and imagery have given farmers numerous new means of improving yields and reducing crop damage. UAV used for practical reasons have taken a peculiar flight in recent decades. New UAV-installed detectors with user's eyes on the earth high-tech cameras and efficient assessment, data collecting, and analysis techniques will be consistently implemented and evaluated. In truth, the utilization of aerial surveillance in the field of agriculture is not recent. For a decade, satellites were employed to survey extensive farmlands and woodland, but with the deployment of UAVs, a new degree of precision and adaptability was

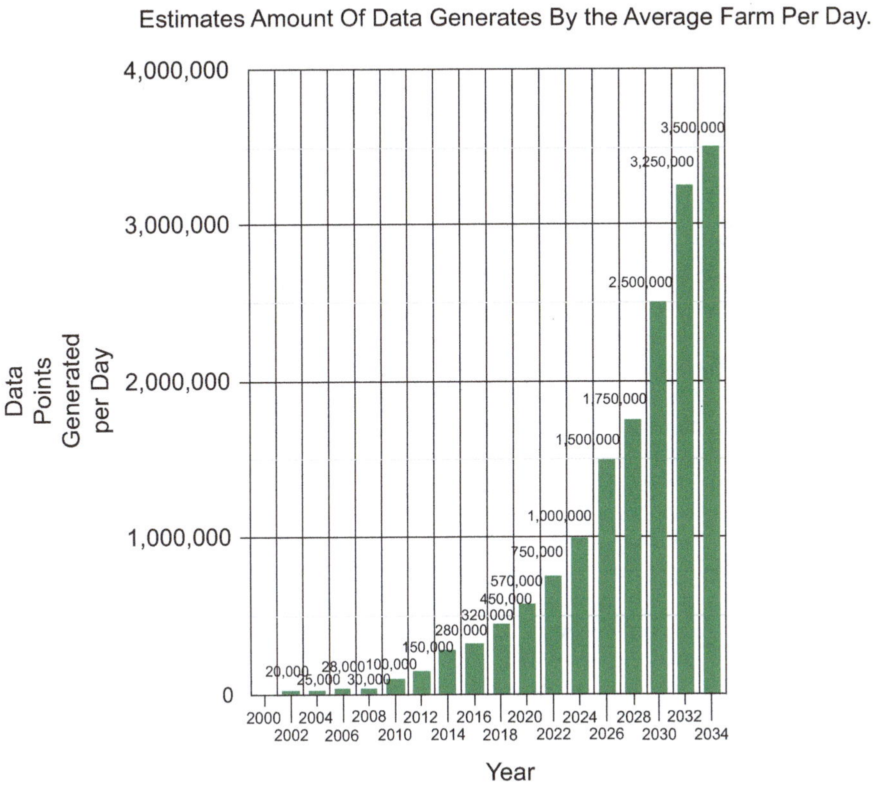

Figure 5: Data generated by average farm per day.

achieved. For UAV flights, it is not necessary to rely on the satellite's location or the right weather conditions. Because UAV photographs are taken from the earth's surface of 400–500 ft, they provide higher quality and reliability (Figure 5).

6 Conclusion

With the development of technology, several industries worldwide have seen a significant transition [38]. Amazingly, while the least digitalized, agriculture has experienced rapid evolvements in developing and marketing agricultural technologies. AI's function in our daily lives has begun to increase our awareness and the power to change our environments [39]. Those new technologies today contribute to multiple sectors by employees limited to only minimum industrial sectors. AI relies on extensive biology, linguistics, information science, arithmetic, and cognitive science and technology. Jha et al. [30] state that a general outline of modern agriculture relies on automation. AI's central notion is for developing a system that works like a human brain [40]. This technology is carried out by understanding in what way the human brain believes and how people learn, decide, and operate while addressing an issue and designing smart design software and systems on this basis.

References

[1] Himesh S. Digital revolution and big data: A new revolution in agriculture. *CAB Reviews: Perspectives in Agriculture, Veterinary Science, Nutrition and Natural Resources*, 13:021, 2018. https://doi.org/10.1079/pavsnnr201813021.

[2] Schimmel Pfennig D, Lowenberg-deboer J. Farm types and precision agriculture adoption: Crops, regions, soil variability, and farm size. *SSRN Electronic Journal*, 2020. https://doi.org/10.2139/ssrn.3689311.

[3] Liakos K, Busato P, Moshou D, Pearson S, Bochtis D. Machine learning in agriculture: A review. *Sensors*, 18(8):2674, 2018. https://doi.org/10.3390/s18082674.

[4] Saiz-Rubio V, Rovira-Más F. From smart farming towards agriculture 5.0: A review on crop data management. *Agronomy*, 10(2):207, 2020.

[5] Kim Y, Evans R, Iversen W. Remote sensing and control of an irrigation system using a distributed wireless sensor network. *IEEE Transactions on Instrumentation and Measurement*, 57(7):1379–1387, 2008. https://doi.org/10.1109/tim.2008.917198.

[6] Dutta PK, Mitra S. Application of agricultural drones and IoT to understand food supply chain during post COVID-19. In: *Agricultural Informatics: Automation Using the IoT and Machine Learning*, 67–87, 2021.

[7] Tzounis A, Katsoulas N, Bartzanas T, Kittas C. Undefined. *Biosystems Engineering*, 164: 31–48, 2017. https://doi.org/10.1016/j.biosystemseng.2017.09.007.

[8] Sarni WM, Kaji JUNKO. From dirt to data, the second green revolution and the internet of things. *Deloitte Review*, 18:4–19, 2016.

[9] Albaji M, Shahnazari A, Behzad M, Naseri A, BoroomandNasab S, Golabi M. Comparison of different irrigation methods based on the parametric evaluation approach in Dosalegh plain: Iran. *Agricultural Water Management*, 97(7):1093–1098, 2010. https://doi.org/10.1016/j.agwat.2010.02.015.

[10] Giles DK, Delwiche MJ, Dodd RB. Undefined. *Transactions of the ASAE*, 30(6):1624–1636, 1987. https://doi.org/10.13031/2013.30614.

[11] Duckett T, Pearson S, Blackmore S, Grieve B. Agricultural robotics: The future of robotic agriculture. UKRAS White Papers, 2018. https://doi.org/10.31256/wp2018.2.

[12] Ferguson RB, Shapiro CA, Hergert GW, Kranz WL, Klocke NL, Krull DH. Nitrogen and irrigation management practices to minimize nitrate leaching from irrigated corn. *Journal of Production Agriculture*, 4(2):186–192, 1991. https://doi.org/10.2134/jpa1991.0186.

[13] Plessen MG, Bemporad A. Shortest path computations under trajectory constraints for ground vehicles within agricultural fields. In *2016 IEEE 19th International Conference on Intelligent Transportation Systems (ITSC)*. 2016. https://doi.org/10.1109/itsc.2016.7795792.

[14] Lee J, Wang J, Crandall D, Sabanovic S, Fox G. Real-time, cloud-based object detection for unmanned aerial vehicles. In 2017 First IEEE International Conference on Robotic Computing (IRC), 2017. https://doi.org/10.1109/irc.2017.77.

[15] Bhaskaranand M, Gibson JD. Low-complexity video encoding for UAV reconnaissance and surveillance. In *2011 – MILCOM 2011 Military Communications Conference*. 2011. https://doi.org/10.1109/milcom.2011.6127543.
[16] Dursun M, Ozden S. A wireless application of drip irrigation automation supported by soil moisture sensors. *Scientific Research and Essays*, 6(7):1573–1582, 2011.
[17] Pedersen SM, Fountas S, Blackmore S. Agricultural robots – applications and economic perspectives. In: *Service Robot Applications*. Intech Open, 2008.
[18] Griepentrog HW, Nørremark M, Nielsen H, Blackmore BS. Seed mapping of sugar beet. *Precision Agriculture*, 6(2):157–165, 2005. https://doi.org/10.1007/s11119-005-1032-5.
[19] Tothill IE. Biosensor's developments and potential applications in the agricultural diagnosis sector. *Computers and Electronics in Agriculture*, 30(1–3):205–218, 2001. https://doi.org/10.1016/s0168-1699(00)00165-4.
[20] Kumar G. *Research Paper on Water Irrigation by Using Wireless Sensor Network*. International Journal of Scientific Research Engineering & Technology (IJSRET), 3–4, 2014.
[21] Shekhar Y, Dagur E, Mishra S, Sankar Narayanan S. Intelligent IoT based automated irrigation system. *International Journal of Applied Engineering Research*, 12(18):7306–7320, 2017.
[22] Huang Y, Hoffmann WC, Lan Y, Wu W, Fritz BK. Development of a spray system for an unmanned aerial vehicle platform. *Applied Engineering in Agriculture*, 25(6):803–809, 2009. https://doi.org/10.13031/2013.29229.
[23] Panpatte S, Ganeshkumar C. Artificial intelligence in agriculture sector: Case study of blue river technology. *Lecture Notes in Networks and Systems*:147–153, 2021. https://doi.org/10.1007/978-981-15-9689-6_17.
[24] Mogili UR, Deepak BB. Review on application of drone systems in precision agriculture. *Procedia Computer Science*, 133:502–509, 2018. https://doi.org/10.1016/j.procs.2018.07.063.
[25] Ahirwar S, Swarnkar R, Bhukya S, Namwade G. Application of drone in agriculture. *International Journal of Current Microbiology and Applied Sciences*, 8(01):2500–2505, 2019. https://doi.org/10.20546/ijcmas.2019.801.264.
[26] Pederi Y, Cheporniuk H. Unmanned aerial vehicles and new technological methods of monitoring and crop protection in precision agriculture. In *2015 IEEE International Conference Actual Problems of Unmanned Aerial Vehicles Developments (APUAVD)*, 2015. https://doi.org/10.1109/apuavd.2015.7346625
[27] Pedersen SM, Fountas S, Blackmore S. Agricultural robots – applications and economic perspectives. In: *Service Robot Applications*. Intech Open, 2008.
[28] Costa FG, Ueyama J, Braun T, Pessin G, Osorio FS, Vargas PA. The use of unmanned aerial vehicles and wireless sensor networks in agricultural applications. In *2012 IEEE International Geoscience and Remote Sensing Symposium*. 2012. https://doi.org/10.1109/igarss.2012.6352477.
[29] Faiçal BS, Freitas H, Gomes PH, Mano LY, Pessin G, De Carvalho AC, Krishnamachari B, Ueyama J. An adaptive approach for UAV-based pesticide spraying in dynamic environments. *Computers and Electronics in Agriculture*, 138:210–223, 2017. https://doi.org/10.1016/j.compag.2017.04.011.
[30] Jha K, Doshi A, Patel P, Shah M. A comprehensive review on automation in agriculture using artificial intelligence. *Artificial Intelligence in Agriculture*, 2:1–12, 2019. https://doi.org/10.1016/j.aiia.2019.05.004.
[31] Kurkute SR. Drones for smart agriculture: A technical report. *International Journal for Research in Applied Science and Engineering Technology*, 6(4):341–346, 2018. https://doi.org/10.22214/ijraset.2018.4061.

[32] Shekhar Y, Dagur E, Mishra S, Sankar Narayanan S. Intelligent IoT based automated irrigation system. *International Journal of Applied Engineering Research*, 12(18):7306–7320, 2017.
[33] Spoorthi S, Shadaksharappa B, Suraj S, Manasa V. Freyr drone: Pesticide/fertilizers spraying drone – an agricultural approach. In *2017 2nd International Conference on Computing and Communications Technologies (ICCCT)*, 2017. https://doi.org/10.1109/iccct2.2017.7972289
[34] Simelli I, Tsagaris A. The use of unmanned aerial systems (UAS) in agriculture. In *7th International Conference on Information and Communication Technologies*. In Agriculture. 730–736, 2015.
[35] Krishna Kagita M. Security and privacy issues for business intelligence in IOT. In *2019 IEEE 12th International Conference on Global Security, Safety and Sustainability (ICGS3)*. 2019, 206–212, 10.1109/ICGS3.2019.8688023.
[36] Kale SD, Khandagale SV, Gaikwad SS, Narve SS, Gangal PV. Agriculture drone for spraying fertilizer and pesticides. *International Journal of Advanced Research in Computer Science and Software Engineering*, 5(12):804–807, 2015.
[37] Huang Y, Reddy KN. Unmanned aerial vehicles: A unique platform for low- altitude remote sensing for crop management. In *Weed Science for Sustainable Agriculture, Environment and Biodiversity: Proceedings of the Plenary and Lead Papers of the 25th Asian-Pacific Weed Science Society Conference*. Indian Society of Weed Science, Madhya Pradesh, India, 185–192, 2015.
[38] Kakkad V, Patel M, Shah M. Biometric authentication and image encryption for image security in cloud framework. *Multiscale and Multidisciplinary Modeling, Experiments and Design*, 2(4):233–248, 2019. https://doi.org/10.1007/s41939-019-00049-y.
[39] Kundalia K, Patel Y, Shah M. Multi-label movie genre detection from a movie poster using knowledge transfer learning. *Augmented Human Research*, 5(1), 2019. https://doi.org/10.1007/s41133-019-0029-y.
[40] Jani K, Chaudhuri M, Patel H, Shah M. Machine learning in films: An approach towards automation in film censoring. *Journal of Data, Information and Management*, 2(1):55–64, 2019. https://doi.org/10.1007/s42488-019-00016-9.

Sarah El Himer, Mariyam Ouaissa, Mariya Ouaissa

General parametric of four hybrid microconcentrator photovoltaic systems for electric car charging station

Abstract: This chapter covers a general, parametric, and performance analysis of four optical elements to be used in microconcentrator photovoltaic systems for electric car charging station using the same primary optic (PO): the Fresnel lens (FL). It concerns two nonimaging optics which are pyramid and hyperbola; and two imaging optics such as circular and square dome.

The study starts from optical modeling of the FL and considers several sizes and geometric needs of secondary optics (SOs). The performance of all concentrator systems is compared in terms of actual concentrations achieved, angles of acceptance, output light power distribution, and tolerances in placement of the secondary element relative to the FL. Results are compared with the same types of SOs but are designed for car roof applications.

For the first simulation results, we found that the FL with 28 prisms and a focal length of 14 mm is the optimal PO, and the pyramid is the most suitable optical element as an SO placed at focal. By adding the silicon solar cell to multijunction solar cell, the efficiency of the optical system is increased by 20%.

By checking if this power can charge an electric car at any power, and by exploiting the roof area of the car, we need 350 silicon cells under 600 multijunction cells. Therefore, the power generated is 50.6 kw. Then, the total surface of our charging station will take 3.5 m^2. By comparing this surface with that of the stations found in Morocco, our station does not take a large surface and with a high power.

Keywords: primary optic, secondary optic, micro-CPV, Fresnel lens, flux uniformity acceptance angle, optical efficiency, pyramid, dome

---

Sarah El Himer, Sidi Mohammed Ben Abdellah University FEZ, Morocco 30000,
e-mail: sarah.elhimer@usmba.ac.ma
Mariyam Ouaissa, Moulay Ismail University, Meknes, Morocco,
e-mail: mariyam.ouaissa@edu.umi.ac.ma
Mariya Ouaissa, Moulay Ismail University, Meknes, Morocco, e-mail: mariya.ouaissa@edu.umi.ac.ma

https://doi.org/10.1515/9783110760477-008

Nomenclature.

A	Exit radius of the SOs	MCPV	Microconcentrator photovoltaic
a_N	Half difference between the input and the output area of the pyramid	n_1	Index of the entrance medium
A_{in}	Entrance aperture area	n_2	Index of the exit medium
A_{out}	Exit or receiver area		
C	Whole concentration ratio	n_s	Index of the secondary optic
C_g	Geometric concentration ratio	N	Number of reflection inside the cone
C_{opt}	Optical concentration ratio	P_{in}	Power at the entrance of the system
		P_{out}	Power at the exit of the system
CPV	Concentrated photovoltaic	PMMA	Poly(methyl methacrylate)
D	Fresnel lens diameter	r'	Solar angle
f	Focal length of Fresnel lens	R	Entrance radius of the Pyramid
FL	Fresnel lens	SO	Secondary optic
H	Length of the SOs	θ_i	SO entrance angle
θ_{out}	Exit angle	θ_f	θ_f
α	Pyramid angle	β	β

1 Introduction

The electric vehicle is the key element of green and ecological transport. It partially solves problems related to energy and the environment [1]. An electric vehicle is a vehicle powered by an electric motor. This motor is powered by a rechargeable battery or by a fuel cell, or even by a thermal generator engine. In fully electric vehicles (battery electric vehicle), the energy sources for this type of vehicle are batteries that can be recharged by the electricity grid or by other renewable energies (solar, wind turbines, etc.), and the photovoltaic shade house (Figure 1) is an autonomous and mobile energy recharging station, for electric vehicles made available to users by local authorities, but also for places for private use, such as shopping centers, vehicle rental companies, or individuals. This is a parking lot for loaded cars. The shade house transforms sunlight into energy, but it protects from its rays and bad weather with a soft and protective shade, among the solar technologies suitable for this application are concentrator photovoltaic (CPV) and micro-CPV (MCPV).

Figure 1: The photovoltaic shade.

CPV is one of the most promising solar-energy-based methods for generating sustainable electricity. It is a high-efficiency replacement for the conventional flat plate module [2]. This method provides for an important reduction in the amount of space required in the cells, the recent halt of activity of several CPV players has posed a challenge to producing a given amount of electrical power while lowering its cost through high efficiency and moderate expenses. This commercial failure contrasts with the technological success that has been attained in the field, with modules nearing 40% efficiency [3].

The goal of employing MCPV is to lower cell size to the sub-mm range in order to increase a variety of benefits while also presenting certain production problems due to the small cell size. The MCPV shrinks CPV modules since their thickness, weight, and volume are related to the cell size; hence, the module's width is reduced.

One or two optical components can be used to make concentration optics. Several writers have utilized various Fresnel lenses (FLs) [4–13], parabolic mirrors [14], and compound parabolic concentrators (CPC) [15, 16]. The results demonstrate that using a single stage for CPV systems cannot meet the general performance criteria; either the system is extremely efficient but the flux distribution on the cell is not homogenous, or the converse is true. Then, multistage concentrators are concentration systems with two optical components: the first of which is known as the primary optic (PO) and the second as the SO which can be more efficient. According to the literature, For a 1,000× desired concentration ratio, we discovered various systems based on the FL, with secondary systems such as the pyramid, the crossed compound parabolic concentrator (CCPC), dome, and hyperbole, which were explored by Rodriguez et al. [17, 18]. The four concentrators have an acceptance angle of less than 1° and an optical efficiency of more than 80%, according to simulation findings. With the pyramid, the best flux uniformity may be achieved. Experiments confirmed these findings, with the optical efficiency falling to 73% and the acceptance angle falling to 0.8°. El Himer et al. [19] evaluated and assessed four CPV systems based on four secondary optical components (CPC, cone, CCPC, and pyramid) connected with a circular flat FL for a 1,000 × as concentration ratio using just simulation. They also demonstrated that, regardless of the material employed, the pyramid as SO provides the most uniform lighting on the solar cell, the widest acceptance angle (1.4°), and the maximum optical efficiency (83%). The performance of parabolic mirrors as a main or SO has been studied by several writers [20, 21]. The results demonstrate that using a parabolic mirror as a major optical element produces intriguing optical results, especially when combined with a flat mirror, as it achieves the highest optical efficiency of 85%. When two optical components of concentration are used, the CPV system is more efficient because it allows for high optical efficiency, high concentration, and a broad acceptance angle [22].

Multiple systems have been investigated in the MCPV system for several years [23, 24], and recent advancements have led to improved microconcentrator ideas and modules [25, 26]. Ritou [25] investigated a system consisting of a 19 × 19 mm^2 conventional Plano convex lenses as the main optic (PO) and a SO, a dome lens with 0.6 mm^2 cells as the SO and a concentration factor of 1,000×. Its findings revealed that the acceptance angle was around 0.7° [26]. With a plano-convex lens POE of 25 mm and a cell aperture of 0.25 mm, design proposed by Jared Hayashi et al. [27] reaches 100 with an acceptance angle of 0.57°. In the Partric research, the diameter of each square lens in the font lens array is 2.0 mm, and each multijunction cell is 0.0632 mm × 0.0632 mm, giving the system a geometric concentration of 0.0632 mm of 1,000×. These results show that the system efficiency is about 92% while the acceptance angle is 0.8° [28]. Salima et al. [29] offered a comparison of two MCPV systems: the first uses a FL as the main optic (PO) and a dome lens as the SO. The second is made up of two lenses. The primary is plano-convex, while the secondary is formed like a dome. Polymethyl methacrylate (PMMA) and polycarbonate (PC) are two potential materials. The results reveal that the system consisting of a PMMA plano-convex lens and a dome has the best optical efficiency in terms of flux nonuniformity; nevertheless, if the two systems are built of PC, they allow for a broad acceptance angle (PC).

This chapter presents the details of the optimization of the primary and four SO used in the MCPV system, this system which will be mounted on a real electric car charging station. The optical performance of the four designed MCPV systems is assessed by ray tracing simulation. Subsequently, the performance of the hybrid MCPV system based on the lens associated with the pyramid is tested.

2 Basic concepts for solar concentrators

Figure 2 depicts a typical optical system that separates two mediums with refractive indices of $n1$ and $n2$ for the entrance and exit environments, respectively.

The system's overall concentration ratio is defined as

$$C = \eta \times C_g \tag{1}$$

where η denotes the optical efficiency given by

$$\eta = \frac{P_{out}}{P_{in}} \tag{2}$$

and C_g denotes the geometrical concentration which is determined by the ratio of the entry and exit surfaces, which is given by:

$$C_g = \frac{A_{in}}{A_{out}} = \frac{n_2 \sin\theta_{out}}{n_1 \sin\theta_{in}} \tag{3}$$

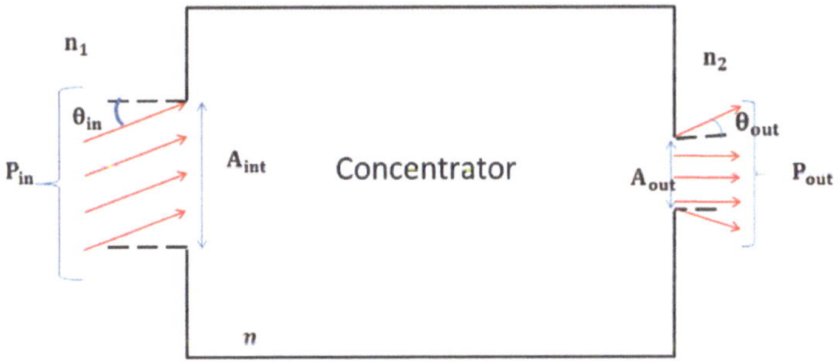

Figure 2: General view of an optical system.

2.1 Nonimaging and imaging optics

V. K. Baranov [30], Martin Ploke [31], and Roland Winston (United States) [32] developed this theory in the mid-1960s in the USSR, Germany, and the United States, respectively, with the study of focusing cones, Martin Ploke [31], and Roland Winston (United States) [32], and it is the most widely used today. Anidolic optics aims to maximize the illumination generated by a source on a target without having to worry about the source producing an image.

Rather than producing an image, one must guarantee that extreme rays traveling in extreme directions and/or emerging from extreme locations at the optical system's entrance aperture are directed to extreme directions and/or extreme positions at the absorber, as illustrated in Figure 3. A lighting source EF, an optical system, and a receiver AB are shown in this diagram. In Figure 3(a), the radiation released from the source's Point F and Point E must be concentrated at Points A and B on the receiver, respectively, and the radiation emitted from the intermediate zones must fall somewhere across the receiver's whole zone AB, due to the optical system's architecture. Alternatively, in Figure 3(b), however, this requirement would always be valid for any point P of the object, in addition to the fact that the light originating from Points F and E must be focused on Points A and B of the picture, respectively. Thus, in the object plan, the distance between the optical axis and the point P is d_O, whereas in the image plan, the distance between the point Q and the optical axis is d_i and are related by the following equation:

$$d_O = M\, d_i \tag{4}$$

with M denoting the optical system's magnification. This produces a picture that correctly reproduces the item by maintaining the relative proportions of its many components.

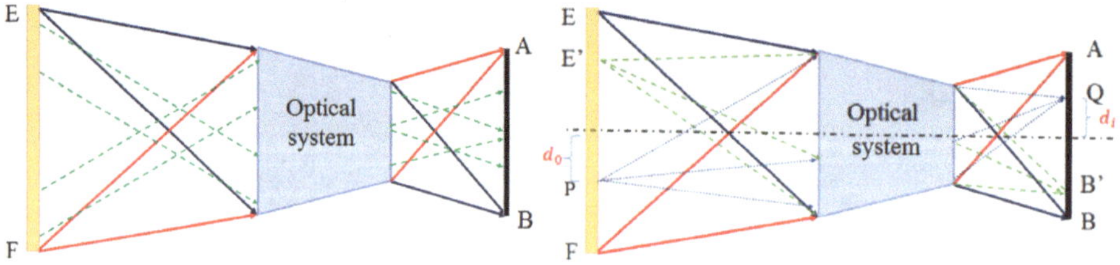

Figure 3: Imaging and nonimaging optical devices propriety. (a) Nonimaging optics; (b) imaging optics.

3 Simulation results

3.1 Optimization of Fresnel lens

The FL is made of a series of narrow concentric grooves. In this part, we will consider the case of a two-dimensional FL with grooves inward. Based on Figure 4, we can draw the following equations:

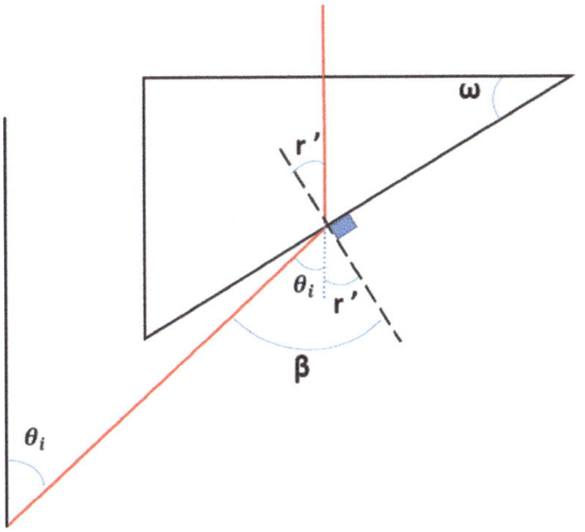

Figure 4: Prism of the Fresnel lens oriented inward.

$$\beta = r' + \theta_\mathrm{i} \tag{5}$$

$$\theta_\mathrm{i} = \mathrm{artg}\frac{d}{2f} \tag{6}$$

$$\tan r' = \frac{d}{2\left(n\sqrt{\left(\frac{d}{2}\right)^2 + f^2 - 1}\right)} \tag{7}$$

The above equations allow us to design the 3D models of the FL with inward grooves shown in Figure 5, where we have taken a FL of 14 mm in diameter, a focal length of 14 mm, and of 14 grooves as an example.

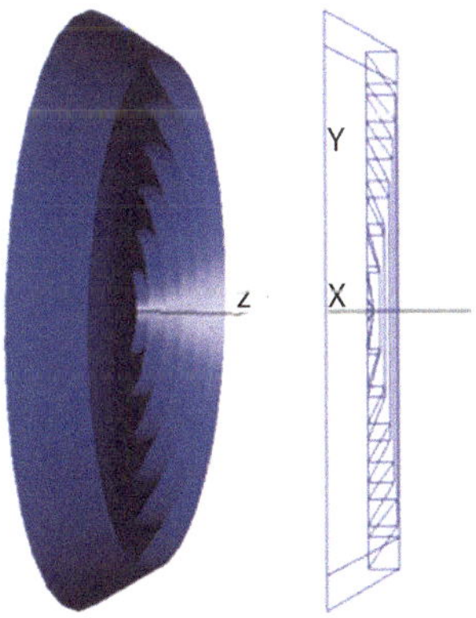

Figure 5: Designed Fresnel lens.

3.1.1 Prism number

Chromatic aberration is caused by the angle at which sunlight refracts from the MCPV module's main optical element depending on the wavelength [32]. Furthermore, due to current limitations in the top cell and restricted semiconductor bandgap energy, infrared photons with wavelengths greater than 1.3 m are not absorbed in either TJ solar cells or silicon sun cells. As a result, when optimizing the SO of the MCPV module, we studied the wavelength range of 0.3 to 1.3 m. We utilized the TracePro (Lambda Research Corporation) raytracing software to assess our MCPV design, which is based on a Monte Carlo technique using the xenon lamp spectrum, and the irradiance used to test the optical efficiency is 1,000 W/m^2. Assuming a low-DNI zone, to generate the three-dimensional model shown throughout the piece, sunlight beams were created using 80% DNI and 20% diffuse light.

In the first test of our FL, we have varied the number of prisms from 7 to 28 to see its effect on the optical efficiency, the angle of acceptance and the flow distribution.

Table 1 resumes the optical performances of the FL proposed one. As illustrated in this table, we found that the FL with 28 prisms could have an optical efficiency that exceeds 87% but it presents the lowest acceptance angle (0.6°); on the other hand, the FL which contains 7 prisms presents the weakest optical efficiency but the widest angle of acceptance (0.8°). Regarding the distribution of the flux, we notice from Figure 6 that whatever the number of prisms, the flux is concentrated with a high intensity in the center of the cell. From 1×10^{11} to 4×10^{10} W/m^2, the intensity drops. It should be noted that this diagram depicts the receiver's center intensity. When it comes to optical efficiency, the more prisms you have, the better. In reality, given a constant diameter lens, reducing the number of prisms increases the angle of the prism. As a result, the prism will be bigger, resulting in losses due to total internal reflection and geometrical losses (losses caused by the divergence of the optical beam) for rays that correspond with the prisms' ends.

Table 1: Performances of Fresnel lens.

	Optical efficiency (%)	Acceptance angle (°)
7 prisms	19.51	0.8
14 prisms	46.14	0.7
28 prisms	87.26	0.6

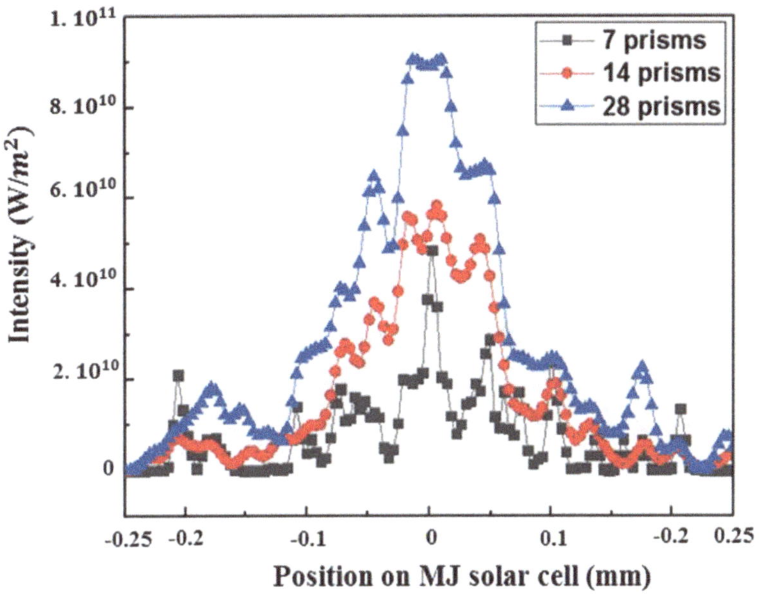

Figure 6: Flux distribution.

3.1.2 Longueur focale

We tested the optical performance of our FL by changing the focal length from 7 to 28 mm in this section. The solar spectrum utilized at this time (visible spectrum) is expanded from 300 to 1,400 nm, and the prism number is set to 28 prisms.

Table 2 resumes the optical performances of the FL as proposed. As illustrated in this table, we found that the FL with 14 and 28 mm as focal length could have an optical efficiency that exceeds 87% but it presents low acceptance angles (0.6° and 0.35°); on the other hand, the FL, which is of 7 mm, presents the weakest optical efficiency but the widest angle of acceptance (2.1°). These results are due to the fact that some light is reflected by totally internal reflection (TIR) and does not reach the receiver at low focal lengths, but at large focal lengths, all rays reach the receiver.

Table 2: Performances of Fresnel lens.

	Optical efficiency (%)	Acceptance angle (°)
7 mm	19.37	2.1
14 mm	87.26	0.6
28 mm	87.8	0.35

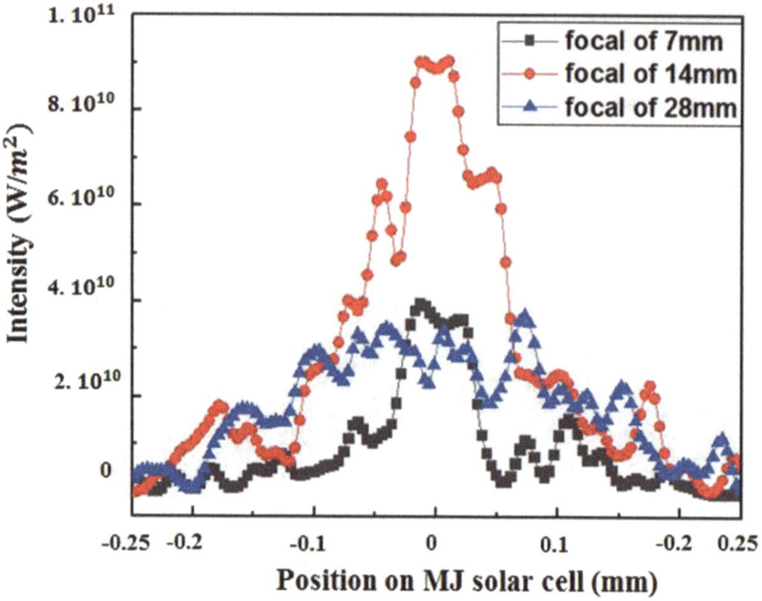

Figure 7: Flux distribution.

The development of the intensity distribution on the receiver is shown in Figure 7 for a variety of focal lengths. In the case of intensity, the intensity reaches a maximum for 14 mm as focal length (*f*-number from 1), and decreases with increasing focal length, but the apparent peak is bigger, showing some homogeneity of the intensity repartition on the receiver.

According to these studies, we have decided to add the SOs to increase the acceptance angle and improved the flux uniformity. We consider for the rest of the work the FL with 28 prisms and a focal length of 14 mm.

3.2 Modeling of the SOs

3.2.1 Nonimaging SO

Here, we proposed two nonimaging SOs which are: the pyramid and the hyperbolic concentrator.

The pyramid (see Figure 8) is a SO. The output and input radius a and R, the output and input angle θ_f and θ_i the angle of the pyramid, and its length H define it. The interior number of ray reflections, n, determines the length of the pyramid:

$$\theta_f = \theta_i + 2n\alpha \tag{8}$$

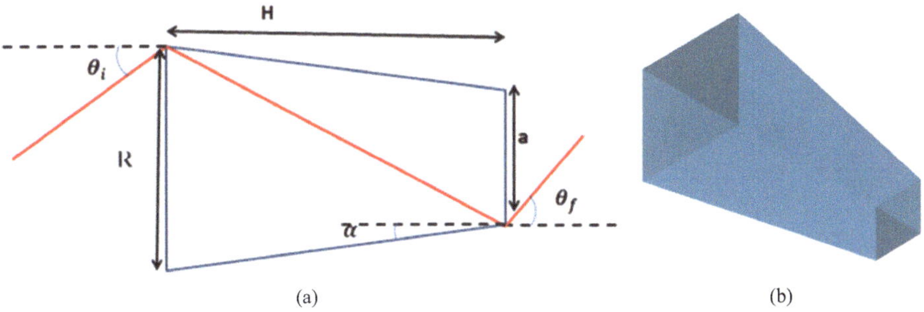

Figure 8: Schematic view of cone/pyramid: (a) 2D view of the cone/pyramid and (b) pyramid in 3D.

$$R = a + \sum_{n=1}^{n_{\max}} a_n \tag{9}$$

$$H = \frac{R - a}{\tan \alpha} \tag{10}$$

More than solar applications, hyperbolic concentrators are utilized for thermal applications. Its design (see Figure 9) was based on equations (11)–(14). The profiles linking the endpoints of the main axis, A and a, and the minor axis, B and b, have the following coordinates:

$$Y_1 = \sqrt{\left(\left(\frac{X}{a}\right)^2 - 1\right) H^2 (CR - 1)^{-1}} \tag{11}$$

$$CR = \left(\frac{A}{a}\right)^2 \tag{12}$$

$$Y_2 = \sqrt{\left(\left(\frac{X}{b}\right)^2 - 1\right) H^2 (CR - 1)^{-1}} \tag{13}$$

$$CR = \left(\frac{B}{b}\right)^2 \tag{14}$$

In this test, we varied the SO heights from 2 to 10 mm when placed on the TJ solar cells. At this point, we fixed the SO entrance at 26° and kept the distance between the FL and the TJ solar cell constant. We computed the irradiance on the TJ solar cell using these SOs. We also analyzed the irradiance by dividing the results into an effective wavelength range (0.3–1.3 µm). Table 3 shows the optical performance of the two SOs with different dimensions. We notice that the optical efficiency for the two SOs increases by increasing the size of the SOs, this is due to the losses of the rays concentrated by the FL, in fact, the surface of the spot of the rays converged by the FL is greater from the entrance surface of the SOs, but from a height of 4 mm,

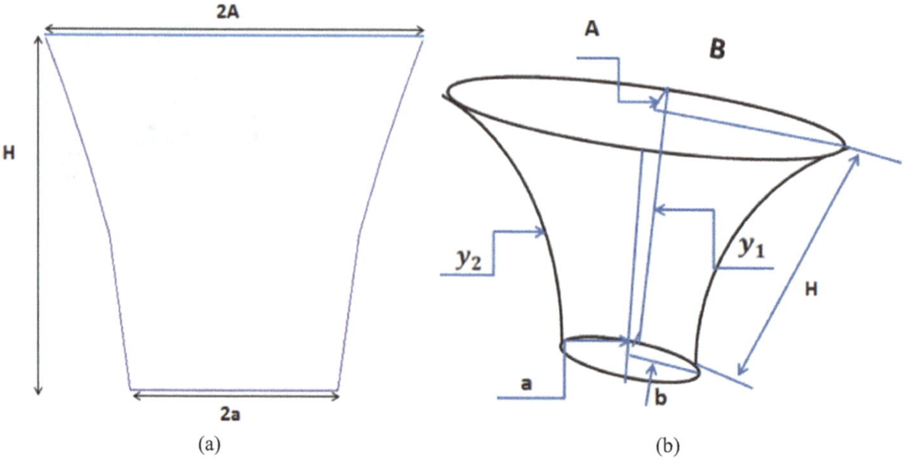

Figure 9: Schematic view of the hyperbole: (a) 2D view of the hyperbole and (b) hyperbole in 3D.

all the rays concentrated by the FL are all picked up by the SOs. Concerning the angle of acceptance, we notice that in the case of the pyramid, the angle of acceptance increases from 0.9° to 1.2°, but in the case of the hyperbola, we find that this angle decreases from 1.6° to 1.2°.

Figure 10(a, b) illustrates the flux distribution on TJ solar cell using the pyramid and hyperbole as SO; in case of using the pyramid, we found that the flux distribution on the TJ solar cell in the effective wavelength range is uniform with an average intensity of $1.5.10^{10}$ W/m^2. In the case of the hyperbole as SO, we can observe the flux distribution is characterized by a very high peak for a height of 2 mm but this peak widens with the increase in the height of the hyperbola until we can have a homogeneous distribution with 10 mm as height.

Table 3: Optical efficiency and acceptance angle of micro-CPV module with different SO heights, as calculated using TracePro.

Heights (mm)	Pyramid		Hyperbole	
	Optical efficiency (%)	Acceptance angle (°)	Optical efficiency (%)	Acceptance angle (°)
2 mm	79.3	0.9	53.23	1.6
4 mm	83.32	1.2	86.26	1.4
6 mm	83.83	1.2	86.33	1.4
8 mm	84.032	1.2	86.42	1.4
10 mm	84.31	1.2	89.26	1.2

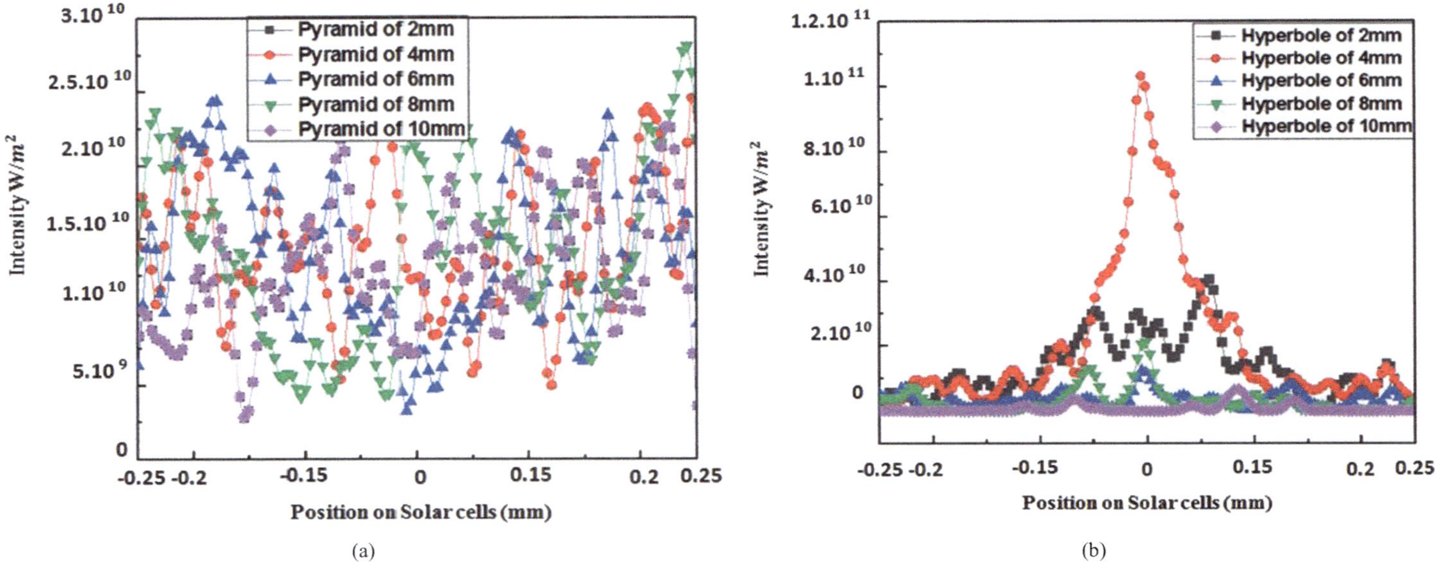

Figure 10: Flux distribution on TJ solar cell: (a) flux distribution on TJ solar cell using a pyramid and (b) flux distribution on TJ solar cell using an hyperbole.

3.2.2 Imaging SO

In this section, we present the dome with square and circular exit as an imaging SO. The objective behind this SO (see Figure 11) is to construct its entry surface in such a manner that every ray coming from the main lens $A(x_1,y_1)$ hits the cell B's opposite extreme point (x_2,y_2). Furthermore, all rays that travel through point A and B, regardless of their angle of incidence, must follow Fermat's principle equation (3) of optical path length conservation [33]. In these circumstances, Fermat's premise is as follows:

$$l_1 + n_{SOE} l_{21} = cst \tag{15}$$

$$l_1 = \sqrt{(x - x_1)^2 + (y - y_1)^2} \tag{16}$$

$$l_2 = \sqrt{(x - x_2)^2 + (y - y_2)^2} \tag{17}$$

The beginning point for constructing the profile is the dome height h_{SO}. It is fixed by forcing the bidimensional profile's tangent to be horizontal at the origin, where $F/\#$ is the main lens's f-number.

$$(x_0, y_0) = (0, h_{SO}) = \frac{x_0}{\tan\left[\arcsin\left(\frac{\sin\left(\operatorname{arctg}\left(\frac{1}{2F/\#}\right)\right)}{n_{SO}}\right)\right]} \tag{18}$$

Table 4 presents the optical efficiency and acceptance angle using circular and square dome. We observe that the high optical efficiency and the large acceptance angle are achieved by the MCPV system using the square dome (86.6%). Concerning the flux distribution illustrated in Figure 12, we can see the two domes that cannot present the uniformity.

3.2.3 Positioning of the SOs from the Fresnel lens

Next, we analyzed the position of the SO from the FL, we consider a SO height of 4 mm, at this size we obtained a wide acceptance angle and high optical efficiency and also to have a compact and light MCPV system. The main idea is to find the optimal position of the SOs regarding the FL.

Following geometrical considerations, three possible positions are a priori possible without any optical losses: The first one, $z = f$, is defined by the focal point of the lens (Figure 13(a)), the second and the third are $z = z_{max}$ and $z = z_{min}$, respectively, defined by the place where the incoming flux size fit with the input diameter (or side) of each element beyond and before the focal point of the lens, as shown on Figure 13(b, c). These positions are defined by

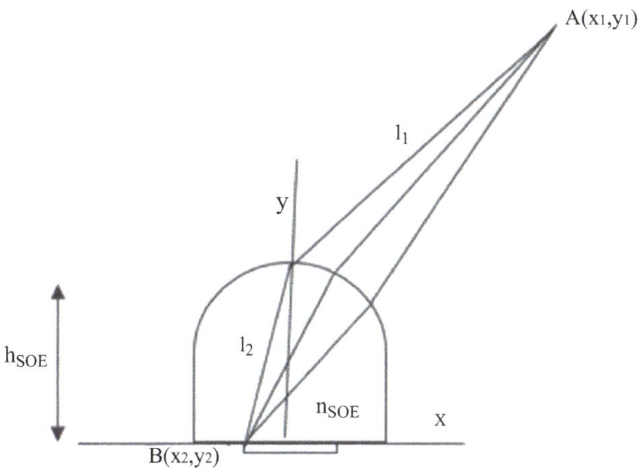

Figure 11: Fermat principle applied to the design of secondary.

Table 4: Optical efficiency and acceptance angle of micro-CPV module with dome as SOs.

	Circular dome	Square dome
Optical efficiency (%)	82.82	86.76
Acceptance angle (°)	0.8	1

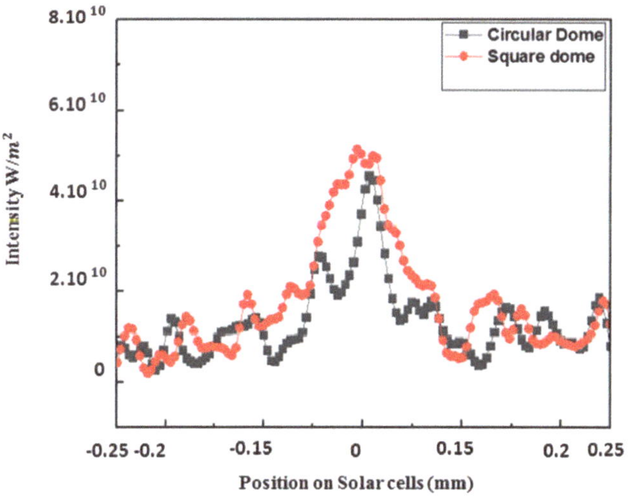

Figure 12: Flux distribution of the dome.

$$Z_{max} = f + \frac{A}{\tan \theta_i} \tag{19}$$

$$Z_{min} = f - \frac{A}{\tan \theta_i} \tag{20}$$

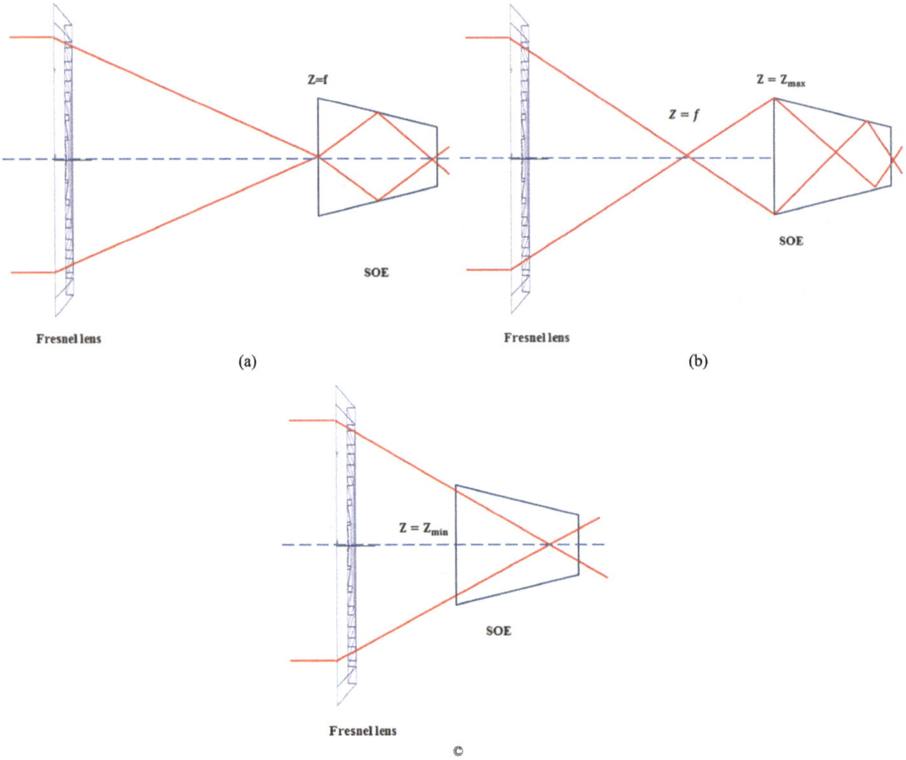

Figure 13: Position of secondary optic: (a) SO at the focal point, z=f; (b) SO at the position maximal, $z = z_{max}$; and (c) SO computed at the minimal position, $z = Z_{min}$.

The changes in optical efficiency of the four systems versus the relative location of the SOs to the lens are shown in Figure 14. It is apparent that the optical efficiency of the hyperbole, circular, and square domes begins to increase and then becomes constant over 2.5 mm in the case of the hyperbole and 1 mm in the case of the two types of domes, then declines as the distance between them increases when approaching to the maximal distance from the lens $z = z_{max}$, concerning the pyramid, we observe that the optical efficiency stills stable from $z = z_{min}$ to $z = z_{max}$. The hyperbole keeps the high optical efficiency.

The acceptance angle of a concentrator is defined as the angle at which the optical efficiency decreases to 80% of its original value [8]. For this, we ran simulations with incidence angles ranging from 0° (normal incidence) to 1.4° (actual most solar tracker precision).

Figure 15 illustrates the changes in the acceptance angle of the complete systems (FL + SO) as the efficiency drops to 80% of its starting value. These findings demonstrate that the exaggeration has the greatest angle (1.4°) for a diameter of 2 mm. In $z = f$, the pyramid reaches 1.2° in all positions, whereas the round and square domes do so in $z = f$.

Figures 16–19 show the flux distribution on TJ solar cell using the four SOs, we observe that the pyramid keep its uniformity for all positions but using the rest of the SOs, we observe a pic with high intensity in the center of TJ solar cell, the high intensity showed when the hyperbole was in focal plan, concerning the circular dome, the high intensity showed in minimal distance.

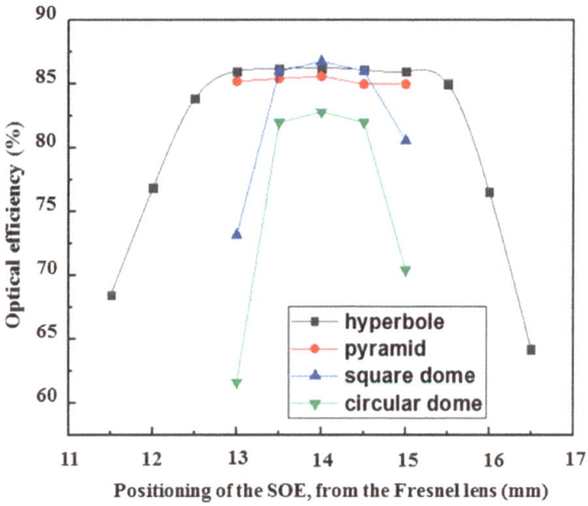

Figure 14: Optical efficiency versus position of secondary optic.

For the rest of the work, we consider a pyramid as an SO placed at the focal plane of the FL, and because at this position, better optical efficiency is obtained with better distribution of the flux and a wide acceptance angle.

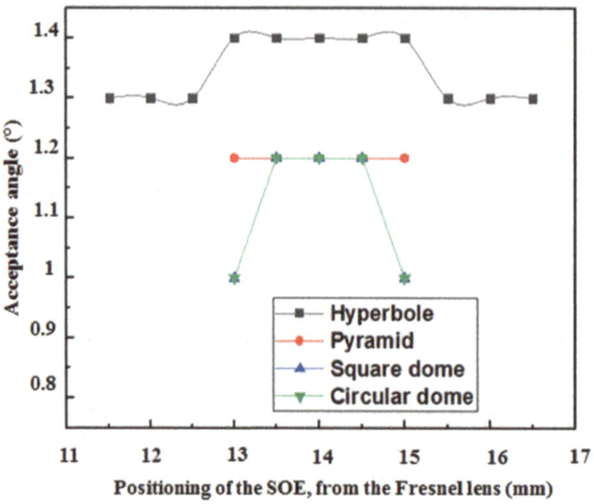

Figure 15: Acceptance angle versus position of the secondary optic.

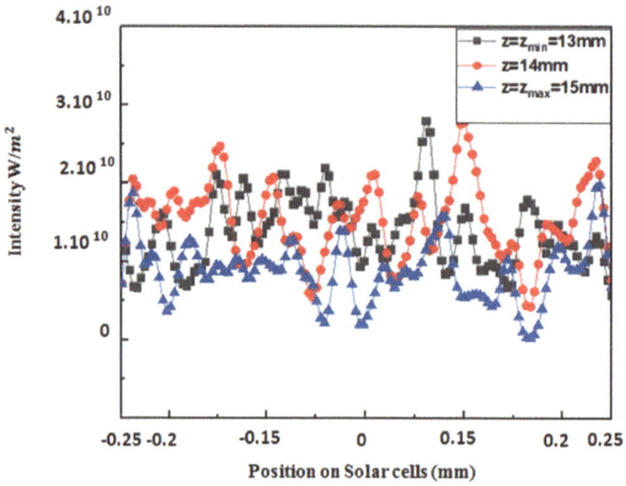

Figure 16: Flux distribution using the pyramid as SO versus position of the secondary optic.

3.3 Hybrid micro-CPV

We put a silicon sun cell of (100 × 100 mm^2) under the MJ solar cell to recover the rays that fall out of the dome and improve the performance of the mini-CPV system, as shown in Figure 20. To evaluate the performance of this hybrid MCPV system, we model it at various incidence angles (ranging from 0° to 20°). Figure 21 depicts the optical efficiency that can be achieved using silicon solar cells. We note

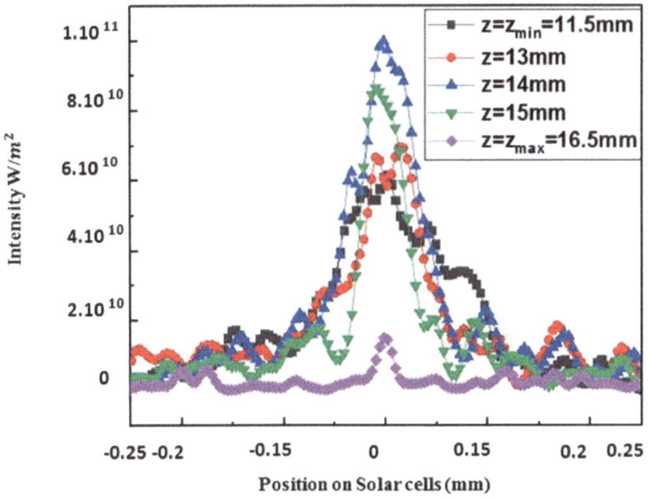

Figure 17: Flux distribution using the hyperbole as SO versus position of the secondary optic.

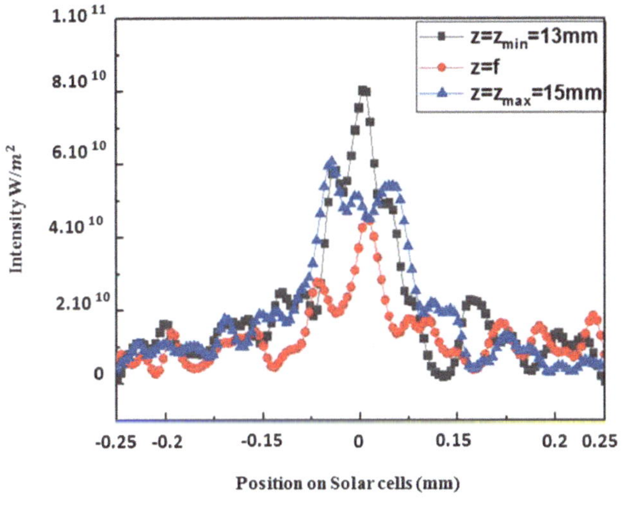

Figure 18: Flux distribution using the circular dome as SO versus position of the secondary optic.

that depending on the cadence angle, diffusion light arriving with inclinations can reach silicon cells as follows:

- For $0° \leq \theta \leq 2°$, the optical efficiency reaches 78% when a portion of the light reaches the pyramid outlet and the rest falls on silicon cells.
- $\theta = 0°$, the TJ solar cell collects all photons at the pyramid output, and the optical efficiency reaches 83%.

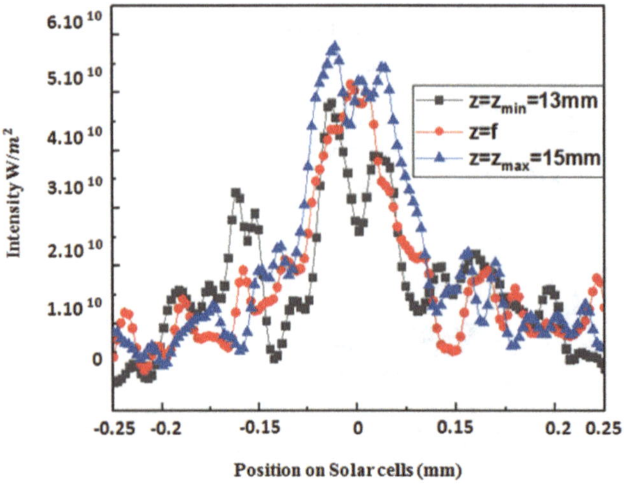

Figure 19: Flux distribution using the square dome as SO versus position of the secondary optic.

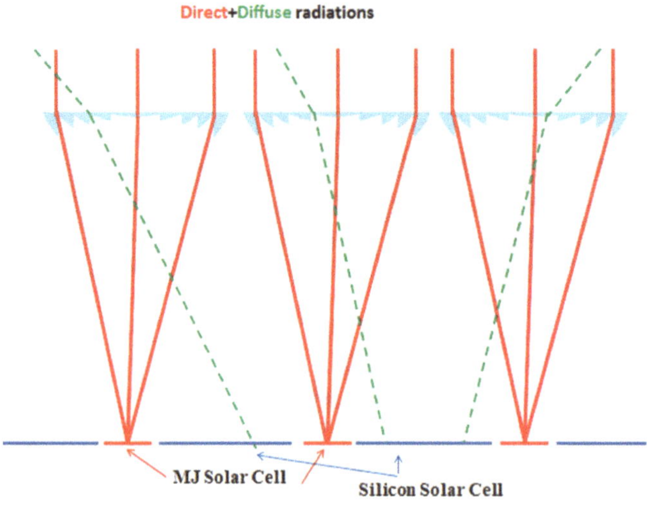

Figure 20: Schematic view of hybrid CPV architecture.

Table 5 shows the comparison of optical performance of the hybrid MCPV to the MCPV system. We can observe that the hybrid MCPV was able to boost the power output by 20%.

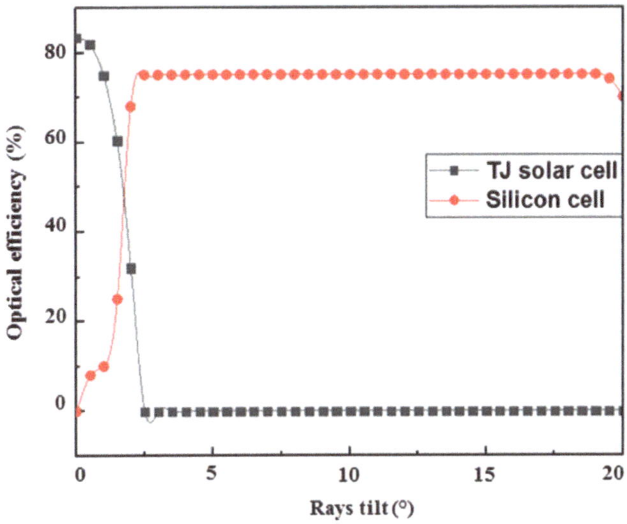

Figure 21: Optical efficiency of the hybrid micro-CPV.

Table 5: Comparison results of optical efficiency.

	Optical efficiency at TJ solar cell				Silicon solar cell			
	(%)		W		(%)		W	
	0°	20°	0°	20°	0°	20°	0°	20
Hybrid micro-CPV	83.32	0	664	0	0	78.22	0	156

4 Charging station proposed

Although an electric car driver rarely comes home with an entirely empty battery, in order to want to recharge it with a solar system dedicated to this use, it would be necessary to be able to produce between 24 and 100 kWh during the day in order to recharge its battery. On a beautiful day, you would therefore have to rely on a 6 kW solar system (24 solar panels measuring 3.5 feet × 5.5 feet) to recharge a car battery.

For an electric car, it is possible to travel about 6.5 km with 1 kWh of energy from the batteries (this figure will vary depending on the type of route and the time of year), which therefore represents 150 Wh/km. Then, in order for an electric car to travel, say, 20,000 km per year, on solar power, it will need to be supplied with about 3,000 kWh of electricity. To produce such an amount of energy, a solar system with a total power of about 50 kW is needed. By checking if this power can charge an electric car at any power, and by exploiting the roof area of the car, we need 350 silicon cells under 600 multijunction cells. Therefore, the power generated is 50.6 kw. Then, the total surface of our charging station will take 3.5 m^2. By comparing this surface with that of the stations found in Morocco, our station does not take a large surface and with a high power.

5 Conclusion

In this study, we have covered a general, parametric and performance analysis of four hybrid MCPV systems for car charging station employing the FL as the main optic (FL). It concerns two nonimaging optics which are pyramid and hyperbola; and two imaging optics such as circular and square dome.

The study starts from optical modeling of the FL and considers several sizes and geometric needs of SOs. All concentrator systems are evaluated in terms of actual concentrations attained, acceptance angles, output light power distribution, and secondary element placement tolerances relative to the FL. The results are compared with the same types of SOs but are designed for car charging station.

For the first simulation results, we found that the FL with 28 prisms and a focal length of 14 mm is the optimal PO, and the pyramid is the most suitable element as a SO placed at focal. By adding the silicon solar cell to multijunction solar cell, the efficiency of the optical system is increased by 20%. By checking if this power can charge an electric car at any power, and by exploiting the roof area of the car, we need 350 silicon cells under 600 multijunction cells. Therefore, the power generated is 50.6 kW. Then, the total surface of our charging station will take 3.5 m^2. By comparing this surface with that of the stations found in Morocco, our station does not take a large surface and with a high power.

References

[1] Mkahl R, Nait Sidi Moh, Wack M. Modeling and simulation of standalone photovoltaic charging stations for electric vehicles. In XIII International Conference on Computer, Communication.

[2] Sanna JC, Gawronska M, Damiano A. Energy Performance of Concentrator and Flat Plate Photovoltaics in the Mediterranean Area. 2015 IEEE International Conference on Renewable Energy Research and Applications (ICRERA), Palermo, Italy, 760–765, 22–25 Nov. 2015.

[3] Steiner M, Siefer G, Schmidt T, Wiesenfarth M, Dimroth F, Bett AW. 43% sunlight to electricity conversion efficiency using CPV. *IEEE Journal of Photovoltaics*, 6(4):1020–1024, Jul. 2016.

[4] El Himer S, Ahaitouf A, El-yahyaoui S, Mechaqrane A. Performance analysis of S. El Himer, A. Ahaitouf, sara El-yahyaoui, abdellah Mechaqrane, Performance analysis of non-imaging Fresnel lens as a primary stage for CPV units, IOP Conference Series: Earth and Environmental Science, 012029, 2018. https://doi.org/10.1088/1755-1315/161/1/012029.

[5] Leutz R, Suzuki A, Akisawa A, Kashiwagi T. Design of a nonimaging Fresnel lens for solar concentrators. *Solar Energy*, 65:379–387, 1999. https://doi.org/10.1016/b978-008043895-5/50228-9.

[6] Akisawa, Hiramatsu M, Ozaki K. Design of dome-shaped non-imaging Fresnel lenses 3 taking chromatic aberration into account. *Solar Energy*, 86:877–885, 2012. https://doi.org/10.1016/j.solener.2011.12.017.

[7] Pham TT, Vu NH, Shin S. Design of curved fresnel lens with high performance creating competitive price concentrator photovoltaic. *Energy Procedia*, 144:16–32, 2018. https://doi.org/10.1016/j.egypro.2018.06.004.

[8] Languy F, Lenaerts C, Loicq J, Thibert T, Habraken S. Performance of solar concentrator made of an achromatic Fresnel doublet measured with a continuous solar simulator and comparison with a singlet. *Solar Energy Materials and Solar Cells*: 70–76, 2013. https://doi.org/10.1016/j.solmat.2012.10.008.

[9] Vallerotto G, Victoria M, Askins S, Antón I, Sala G. Experimental characterization of achromatic doublet on glass (ADG) Fresnel lenses. *AIP Conference Proceedings*, 1881:030010, 2017. https://doi.org/10.1063/1.5001421.

[10] Zhuang Z, Yu F. Optimization design of hybrid Fresnel-based concentrator for generating uniformity irradiance with the broad solar spectrum. *Optics & Laser Technologys*, 60:27, 2014. https://doi.org/10.1016/j.optlastec.2013.12.021.

[11] Wang G, Chen Z, Hu P, Cheng XF. Design and optical analysis of the band-focus Fresnel lens solar concentrator. *Applied Thermal Engineering*, 102:695–700, 2016. https://doi.org/10.1016/j.applthermaleng.2016.04.030.

[12] Ryu K, Rhee JG, Parkc KM, Kim J. Concept and design of modular Fresnel lenses for concentration solar PV system. *Solar Energy*, 80:1580–1587, 2006. https://doi.org/10.1016/j.solener.2005.12.006.

[13] Pham TT, Vu NH, Shin S. Novel design of primary optical elements based on a linear Fresnel lens for concentrator photovoltaic technology. *Energies*, 12:1209, 2019. https://doi.org/10.3390/en12071209.

[14] Chong KK, Siaw FL, Wong CW, Wong GS. Design and construction of non-imaging planar concentrator for7 concentrator photovoltaic system. *Renewable Energy*, 34:1364–1370, 2009. https://doi.org/10.1016/j.renene.2008.09.001.

[15] Tian M, Su Y, Zheng H, Pei G, Li G, Riffat S. A review on the recent research progress in the compound parabolic concentrator (CPC) for solar energy applications. *Renewable and Sustainable Energy Reviews*, 82:1272–1296, 2018. https://doi.org/10.1016/j.rser.2017.09.050.

[16] Yu Y, Liu N, Li G, Tang R. Performance comparison of CPCs with and without exit angle restriction for concentrating radiation on solar cells, angle restriction for concentrating radiation on solar cells. *Applied Energy*, 155: 284–293, 2015. https://doi.org/10.1016/j.apenergy.2015.06.012.

[17] Rodríguez JPF, Baig H, Fernández EF, Almonacid F, Mallick T, Higueras PP. Optical modeling of four Fresnel-based high-CPV units. *Solar Energy*, 155: 805–815, 2017. https://doi.org/10.1016/j.solener.2017.07.027.

[18] Rodríguez JPF, Fernández EF, Baig H, Almonacid F, Mallick T, Higueras PP. Development, indoor characterisation and comparison to optical modelling of four Fresnel-based high-CPV units equipped with refractive secondary optics. *Solar Energy Materials and Solar Cells*, 186: 273–283, 2018. https://doi.org/10.1016/j.solmat.2018.06.050.

[19] El Himer S, Ahaitouf A, El-yahyaoui S, Mechaqrane A. Parametric optimization and performances analysis of four secondary optical elements for concentrator photovoltaic systems. *International Journal of Renewable Energy Research-IJRER*, 8:4, 2018. https://doi.org/10.1063/1.5053502.

[20] Lokeswaran S, Mallick TK, Reddy KS. Design and analysis of dense array CPV receiver for square parabolic dish system with CPC array as secondary concentrator. *Solar Energy*, 199: 782–795, 2020. https://doi.org/10.1016/j.solener.2020.0.

[21] Trespidi F, Timò G. Optical design of an off-axis mirror based solar concentrator. *AIP Conference Proceedings*, 1881:030008, 2017. https://doi.org/10.1063/1.5001419.

[22] Trespidi F, Malvisi E, Parmesani R. Shape characterization of the primary mirror of a mirror based solar concentrator. *AIP Conference Proceedings*, 1881:030009, 2017. https://doi.org/10.1063/1.5001420.

[23] Rodríguez JPF, Fernández EF, Solas ÁF, Almonacid F, Talavera DL, Higueras PP. Optimization of an ultra-high CPV Cassegrain-Koehler unit with 2000_ concentration ratio. *AIP Conference Proceedings*, 2149: 70004, 2019. https://doi.org/10.1016/j.enbuild.2017.01.048.

[24] Cruz-Campa JL, et al. Microlens rapid prototyping technique with capability for wide variation in lens diameter and focal length. *Microelectronics Engineering*, 87(11):2376–2381, 2010.

[25] Ritou A. Développement, fabrication et caractérisation de modules photovoltaïques à concentration à ultra haut rendement à base de micro-concentrateurs, PhD thesis Grenoble Alpes University, 2016.

[26] Yoon J, et al. Ultrathin silicon solar microcells for semitransparent, mechanically flexible and microconcentrator module designs. *Nature Materials*, 7(11):907–915, 2008.

[27] Jared BH, et al. Micro-concentrators for a microsystems-enabled photovoltaic system. *Optical Society of America*, 22, 2014.

[28] Domínguez C, et al. A review of the promises and challenges of micro-concentrator photovoltaics. *AIP Conference Proceedings*, 1881, 2017.

[29] Ayane SEL, Himer SE, El-Yahyaoui S, Ahaitouf A. Performance Analysis of a Two Stage Micro Photovoltaic Concentrator, Proceedings of 2019 7th International Renewable and Sustainable Energy Conference, IRSEC 2019.

[30] Baranov VK. Parabolo-toroidal mirrors as elements of solar energy concentrators. *Applied Solar Energy (USSR) (Engl. Transl.) (U. S.)*, 3:11–14, 1996.

[31] Ploke M. Lichtführung Einrichtungen mit starker Konzentrationswirkung. *Optik*, 25, 31.7, 1967.

[32] Hinterberger H, Winston R. Efficient light coupler for threshold Cˇ erenkov Counters. *AIP Review of Scientific Instrument*, 37:1094–1095, 1966.

Siham Eddamiri, Meriem Houmer, Mariya Ouaissa,
Mariyam Ouaissa

The influence of social media in improving supply chain intelligence using deep learning

Abstract: In recent years, there has been a significant rise in the analysis of consumer reviews expressed on social media, which can be used to obtain more meaningful and relevant information via market study to predict consumer behavior. As a result, sentiment analysis plays a crucial role in evaluating customer satisfaction and identifying consumer opinions, attitudes, and behaviors toward certain topics. In this chapter, we present a system for sentimental classification of reviews posted on social media in a way that can provide an advantage for supply chain managers. More clearly, the proposed system is a hybrid method of deep learning architectures, specifically recurrent neural networks and long short-term memory, using a word embedding word2vec to embed the features in a review to predict the sentiment. Our system achieved promising results which demonstrate the effectiveness of our approach.

Keywords: deep learning, supply chain management, social network analysis, sentiment analysis, word embedding, LSTM

1 Introduction

In the world of social media, the web 2.0 revolution is transforming the world. Not only individuals but also businesses and decision makers may interact and enhance their goods and services by connecting with other companies through social media networks. Moreover, with the emergence of the new technology revolution and economic globalization, as well as high customer expectations, companies' supply chain management (SCM) has changed significantly, emphasizing the significance of competition among supply chains rather than companies [1]. Hence, modern SCM systems must be able to connect to numerous networks [2].

Siham Eddamiri, Department of Mathematics and Computer Science, University Moulay Ismail, ENSAM, Meknes, Morocco, e-mail: sihameddamiri@gmail.com
Meriem Houmer, University Moulay Ismail, Meknes, Morocco, e-mail: m.houmer@gmail.com
Mariya Ouaissa, University Moulay Ismail, Meknes, Morocco, e-mail: mariya.ouaissa@edu.umi.ac.ma
Mariyam Ouaissa, University Moulay Ismail, Meknes, Morocco, e-mail: mariyam.ouaissa@edu.umi.ac.ma

https://doi.org/10.1515/9783110760477-009

Social media and networks have undoubtedly gained a lot of attention from big data researchers in many applications [3]. The advent of social media makes communication between producers, consumers easier and more convenient. A massive amount of data are generated every time a user communicates with another on a social platform such as Twitter, Facebook, and Myspace [4]. Many different types of media, such as text, videos, pictures, and music are published and shared on social media [5]. This paves way for a better understanding of people's behavior from this type of information. When it comes to analyzing and understanding the communication happening across transcripts, sentiment analysis (SA) or opinion mining, and natural language processing (NLP) play a vital role.

There have been significant advances in the field of SA since the early 2000s. Many different terms are used for SA, including opinion mining, sentiment classification, and opinion extraction. From text, it can be used to identify favorable or negative feelings toward things like products or services or companies or people [6] by converting them into positive, negative, or natural sentiment [7]. SA was largely employed in most of the current research studies to help comprehend their consumers and make required judgments to improve their products or services as a consequence [8]. Social media sentiments can certainly be used to develop business strategies. It is often used to manage a product or brand's online reputation. In addition, by analyzing large amounts of customer data, it is possible to identify flaws in goods or services, study consumer perceptions of the company and its products, uncover latent customer needs, and improve corporate operations in general. Data analysis will enable companies the potential to properly predict future trends and make more intelligent decisions in numerous SCM areas by evaluating vast volumes of data.

The proposed approach is based on developing a framework that uses a hybrid approach for the sentiment classification of reviews posted on social media in managing different areas of the supply chain. This is accomplished by (1) doing preprocessing; (2) generating semantic representations of customer opinion using an recurrent neural network (RNN)–long short term memory (LSTM) model; (3) using various classifier algorithms to categorize sentiment as positive or negative; and (4) analyzing the results. More clearly, positive requests express interest or appreciation, while negative requests express regret and disapproval.

The remainder of this chapter is organized as follows: Section 2, we look at related works. In Section 3, we go into further depth about our strategy. By analyzing our system set to alternative approaches, we offer an experimental assessment in Section 4. Conclusion and future efforts are discussed.

2 Related works

2.1 Supply chain

Social and technological variables have a major influence on a supply chain's complexity. On a technical level, it is a system that involves a big number of retailers and distributors and vendors and customers and producers and production plants [9]. There has been a strong focus on supply chain modeling and other "hard" topics related to incorporating the latest technology in this area. According to Shub and Stonebraker [10], "soft" human resource activities and organizational variables have not been given enough attention when it comes to improving supply chain strategies and success. To the contrary, despite several specialists highlighting the importance of trust and mutual understanding within supply chains, social aspects of SCM are not well studied [11]. There is also a change in attention away from improving the quality of the supply chain by strengthening ties among suppliers as chances for supply chain performance advancement via advanced technologies [12]. It is therefore possible to make a profit by studying the relationship between social media content and supply chain functions [13]. A supply chain using social media can help discover patterns and make better business decisions [3], such as learning about industry rivalry, establishing customer relationships, or finding key performance metrics using input from customers [14]. Through big data analytics, businesses will be able to make better decisions in numerous aspects of SCM. Therefore, there is an urgent need for artificial intelligence methods to cope with this big data-related challenge and to use them in different aspects of supply chains, as most literature researches have only considered one or two areas of the supply chain [15]. For instance, several authors use applied machine learning (ML) techniques to design an automated SCM framework such as predict supply chain risks by Baryannis [16] or predict financial risks in supply chains [17] or even segment suppliers based on environmental factors using ML hybrid approach [18].

2.2 Sentiment analysis

Since the booming of social media platforms has taken over the communication world, it is crucial that the data residing on these platforms provide useful insight into the opinion, mood, and sentiment of people regarding any product or idea. However, there are several automated techniques that have been developed to detect opinions in social media-generated free text because manual processing of such humongous data is prohibitively expensive. This method is referred to as SA in the context of this article. It involves the analysis of social media users' opinions on companies and goods. The SA is a method of assessing a document or review to identify if the reviewer has a positive or negative or even neutral attitude [6, 19].

To date, SA has been successfully applied to a wide range of fields, including NLP, statistics, and ML. As a part of SA, polarity or orientation is used to determine and record the public's view on specific issues in text [20]. In fact, SA is a crucial tool for optimizing a large component of customer satisfaction management in today's competitive market [21]. As a result, SA research has grown in the last two decades, with numerous applications to business intelligence, such as stock price prediction, ticket sales forecasts, and customer retention evaluation. For instance, Tang et al. [22] assess the semantic orientation of words or sentences using lexicon-based techniques. Once the text is represented by a bag of words model, it is categorized using a variety of statistical ML techniques. Prediction rules are created by using statistical methodologies to classify training data, which is an ML process [23, 24].

3 Methodology

Our system is described in detail in this section, which uses a hybrid approach of deep learning architectures with word embedding for SA. Figure 1 illustrates the classifier's general architecture. There are several steps to this system: clean the noisy social media reviews. Then, for each review, create word embeddings using the skip-gram (SG) model. Reviewer's opinion representations are then obtained as feature vectors using the LSTM-RNN model. Next, classify them into three categories positive, negative, or neutral by using several classifiers. Finally, identify the sentiment of the consumers.

3.1 Text preprocessing

Social media reviews are fairly frequently expressed in language that is full of unusual characters, symbols, hyperlinks, and so on. As a result, we eliminate the noise from the dataset using regular expressions. Prior operations on the input text are performed in the preprocessing step in order to convert them into a format that the classifier can accept. As a result, all reviews are preprocessed into tokens of words. Once the duplicate words have been removed, each word is given a unique representation. In the end, a word index and key are utilized to build the vocabulary dictionary. Next, for each review, the feature vectorization process has been performed after preprocessing to generate embedding social reviews.

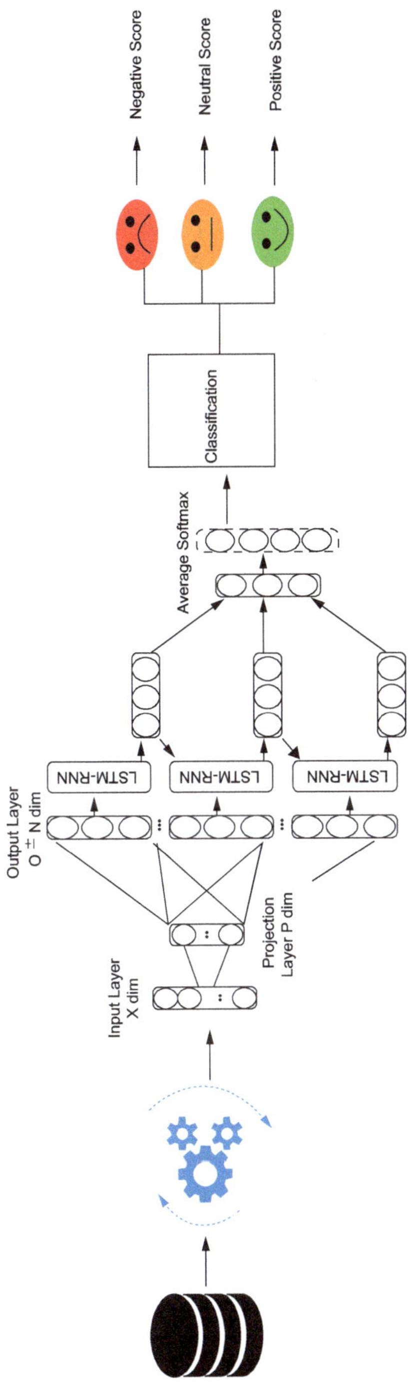

Figure 1: Schematic process of the proposed approach.

3.2 Word embedding

It has recently been discovered that word embedding can be used as a means of assigning probability values to target words inside text (text documents). Since these models are based on customer opinions, our initial aim is to alter them based on those opinions. With the neural network models built before, the second goal is creating numerical values for each item in a latent feature space. In the field of neural language modeling, word2vec is one of the most commonly used and effective models. A model called word2vec was developed in 2013 by Mikolov at Google [25]. The SG model is used to anticipate context words based on the target words in a sentence or paragraph. An average log probability of $w_1, w_2, w_3, \ldots, w_T$ and a context window of size c is the goal of this model, as shown in Figure 2.

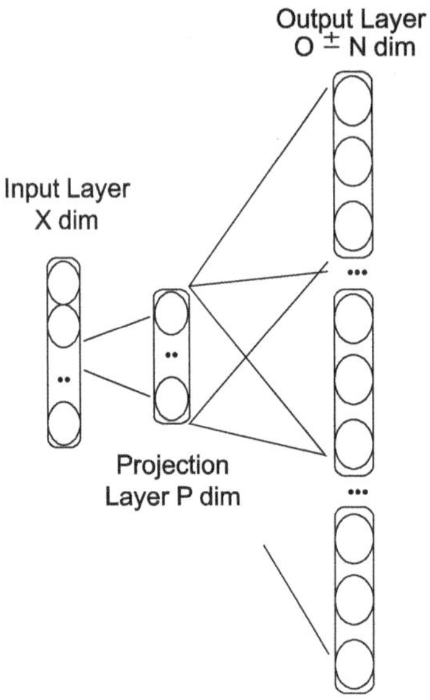

Figure 2: The skip-gram model architecture.

3.3 LSTM-RNN for review representation

For each of the social media reviews, a SG graph offers a Boolean array (0 or 1) representation. Moreover, a further problem arises from the enormous dimensionality of the bag of words function. A neural network model (word2vec) is suggested to represent the semantic connection between words. A sequence of words is not considered by word2vec, but just the semantics of each individual word. The sequence of terms, their syntax, and the semantic link between them all need to be improved. Social media reviews are presented in a new light in this perspective. Figure 3 shows a

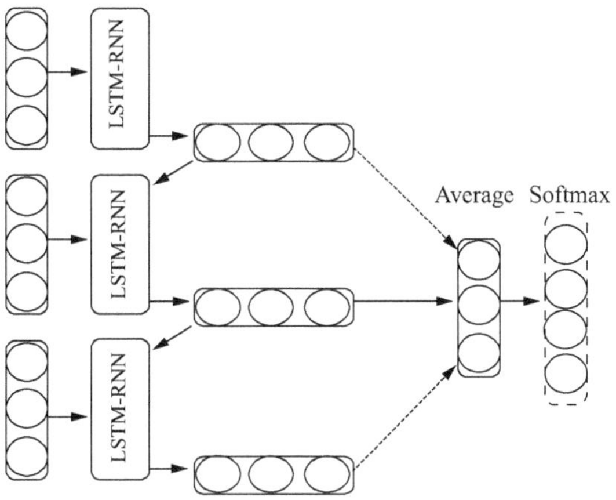

Figure 3: Process of extracting vector representation of social media reviews.

high-level perspective of the social media reviews data. These are RNNs that Hochreiter first introduced as a kind of LSTM. To eliminate long-term dependencies, LSTMs have been designed with unique architecture. There are three gates: the input gate, output gate, and forget gate that make up an LSTM's normal design. Structure of deep representation requires extraction of SG using predicate repository, learning of $|V|$ spatial representation word2vec using $|U$ spatial representation, and learning of deep representation cells (LSTM cells) using $|S|$ spatial representation, all of which require $|U|$ spatial representation. In fact, $|D|$ spatial representation of the provided reviews is returned by this procedure. Each of the three dimensions of the representation has a sequence network (RNN) with n hidden units $(h = h_1, h_2, \ldots, h_n)$ that are hidden from view in the hidden layer. It accepts a $|V|$ dimensional representation $(y = y_1, y_2, \ldots, y_n)$ as input and outputs a $|D|$ spatial representation $(z = z_1, z_2, \ldots, z_n)$. On average, the previous state s_{i-1} and a term y_i are transformed each time s_i to produce the next state z_i and output word s_i. Moreover, s_i represents the neighbors of the ith word in h, whereas z_n represents the review output from h_n. This is because every hidden unit in a RNN applies the function f again as the following formula:

$$f:\{s_{i-1}, y_i\} \rightarrow \{s_i, z_i\}$$

When given a set of word vectors, LSTM-RNNs create their respective vertex vectors. Vertices are analyzed in terms of their neighbors, a series of nodes vertices and edges. Each hidden state h_t is averaged to represent a vertex vector. The following steps are taken by our system to create a social media review vector. Utilizing the SG model, we construct the words embeddings $w_1, w_2, \ldots, w_t, \ldots, w_n$ for a particular node v_r. The LSTM-RNN model uses these word representations as input.

3.4 Classification of sentiment reviews

ML is the process of predicting and recognizing patterns and then generating appropriate results based on that knowledge. ML algorithms analyze data patterns and learn from them. An ML model will learn from each attempt and improve. Data must be split into training and test sets before a model can be evaluated. Next, a selection of performance metrics needed to be added to our model's classification of sentiment reviews. In this case, we attempted to classify reviews by positive, negative, or neutral opinion. We used five algorithms for our modeling purpose: decision tree (DT), random forest (RF), logistic regression (LR), K-nearest neighbor (KNN), and support vector machine (SVM).

- **KNN classifier** is an easy-to-use method that works well for a wide range of problems and datasets. With positive integer K and a test observation N_0, KNN classifier first calculates the distance and identifies K points (neighbors) in training data that are closest to test observation x_0 Next, it calculates the conditional probability for class j as the fraction of points in N_0 whose response values equal class j:

$$\Pr(Y=j|X=x_0)=\frac{1}{k}\sum_{i\,\mathrm{in}\,N_0} I(y_i=j)$$

  In the end, KNN classifies x_0 according to the Bayes rule.
- **DT classifier** is the tree structure of the DT (which can be a binary tree or a nonbinary tree). Its nonleaf nodes represent features being tested, its branches the output of an attribute throughout a range of values, and its leaf nodes a category value. To reach the leaf node, an output branch is chosen under its value. Therefore, the leaf node's decision result is saved as a category.
- **RF classifier** is a supervised learning algorithm that utilizes trees as basic components in the construction of prediction models that are more accurate and powerful. These DTs are unrelated to one another. It was determined by the algorithm parameter d how many levels each branch of the DT should have. The Gini index for this attribute is calculated as

$$G(X_i)=1-\sum_{j=1}^{J}\Pr(X_i=L_j)^2$$

  In this case, G is the Gini index, and the higher the value of G, the greater the degree of uncertainty in data.
- **LR classifier** is an attribute control model based on advanced linear regression. Its output value range is limited to [0, 1] due to the employment of a sigmoid function. Explanatory variables and susceptibility to landslides:

$$f(z) = \frac{1}{1 + \exp -z}$$

where $z = w_1x_1 + w_2x_2 + \cdots + w_Mx_M + b$ denotes a weighted combination model, and b represents the function's intercept; w_i specifies the function's correlation coefficient.

- **SVM classifier** is a supervised ML algorithm that uses statistical learning theory and the structural risk minimization concept. For nonlinear universe, a hyperplane is used to reformat it. There are just two possible values for the labels of an SVM class, namely +1 or −1:

$$f(x) = \text{sign}(w, x) + b$$

Where a training dataset $S = (x_1, y_1), \ldots, (x_n, y_n)$ and x_i in R^n and $y + 1, -1$. Finding the hyperplane that yields optimum separation between training samples is what drives this approach.

4 Evaluation

4.1 Dataset

As part of this study, the system was evaluated using two separate datasets from diverse areas such as movie reviews and airline reviews. Since they are all from different fields, their writing styles are different. There may be a variety of word dependencies in a review. An independent SA model was created by combining the training data from each dataset and evaluating it independently. These datasets all have the same number of positive, negative, and neutral samples or are extremely close to having the same number of samples. The datasets are discussed in the following way:

- **Movie review:** in this dataset, there are 25,000 highly polarized moving reviews (good or bad) for training and the same amount again for testing in the IMDB dataset collected by Stanford researchers and was used in a 2011 paper [26]. When it comes to describing movie reviews, it is difficult to tell if they are positive or bad. Using an 80% in training and 20% in the testing of the data.
- **Airline review:** the source data for this dataset was obtained from Crowdflower's Data for Everyone repository. A review of various US airlines is included in the book. In February 2013, they collected Twitter data from February 2013 to January 2014 and then categorized it into positive, negative, and neutral tweets [27]. Then, they divided the reasons for the bad tweets into categories (such as a late flight or rude service). On the right-hand side, you can see if the tweets were positive, neutral, or negative.

4.2 Validation metrics

With the use of three performance evaluation metrics (accuracy, AUC, and ROC) for classifier comparison, we will determine the efficacy of our methods.

- **Accuracy** is the ratio of the number of correct predictions by the classifier to the total number of samples for a given test dataset. The formula is as follows:

$$\text{Accuracy} = \frac{\text{TP} + \text{TN}}{\text{TP} + \text{TN} + \text{FP} + \text{FN}} \tag{1}$$

- ***F*1-score**, as well as being known as a balanced *F*-score, is described as the balanced combination of precision and recall:

$$F1 - \text{score} = 2 * \frac{\text{precision} * \text{recall}}{\text{precision} + \text{recall}} \tag{2}$$

- **Receiver operating characteristic curve (ROC)** is one of the most well-known graphs, which shows how well a binary classifier performs. By charting the true positive rate versus the false positive rate at various threshold settings, the curve is formed. Because of this, a curve that is closer to the top of the left side of the graph will have a better prediction:

$$\text{FPR} = \frac{\text{FP}}{\text{FP} + \text{TN}} \tag{3}$$

$$\text{TPR} = \frac{\text{TP}}{\text{TP} + \text{FN}} \tag{4}$$

- **AUC (area under the curve) value** is a metric for binary classification that measures the accuracy of model, ranging from 0.5 to 1. Instead, it measures alternative discrimination capacity based on the ROC curve's receiver operating characteristic (AUC). Assume that the points $(x_1, y_1), (x_2, y_2), (x_3, y_3), \ldots, (x_n, y_n)$ are connected in a ROC curve. This can be calculated as follows:

$$\text{AUC} = \frac{1}{2}\sum_{i=1}^{n-1}(x_{i+1} + x_i).(y_{i+1} + y_i) \tag{5}$$

4.3 Experimental process

In this section, using data preparation as a preliminary step, we evaluate our methodology process for analyzing consumer sentiment. As soon as the features have been scaled and selected, the learning difficulty can be reduced by prioritizing the features that are most important to the aim. Finally, we use several classifiers to classify the review of the consumers into three categories positive, negative, or neutral. More clearly, after preprocessing the data social reviews, a new data is generated without the noises data. Next, we generate a vector embedding for each review given by each consumer using the LSTM-RNN with word2vec. Our suggested system is compared to common classifiers such as DT, RF, LR, KNN, and SVM. The performance of these algorithms is presented in Tables 1, 2, and 4. When comparing these algorithms, we can see that the RF method outperforms them in terms of accuracy.

Table 1: Evaluation metrics comparison of the different techniques in the IMDB dataset.

Models	Accuracy (%)	AUC	*F*1-score	Recall
Our system	98%	0.98	0.97	0.97
Random forest	96%	0.95	0.97	0.93
KNN	79%	0.78	0.91	0.90
Decision tree	94%	0.95	0.95	0.92
SVM	73%	0.75	0.81	0.67
Logistic regression	70%	0.76	0.80	0.81

Table 2: Evaluation metrics comparison of the different techniques in the airline dataset.

Models	Accuracy (%)	AUC	*F*1-score	Recall
Our system	99%	0.99	0.98	0.97
Random forest	97%	0.94	0.96	0.94
KNN	81%	0.79	0.93	0.91
Decision tree	95%	0.97	0.96	0.94
SVM	79%	0.77	0.90	0.72
Logistic regression	78%	0.79	0.85	0.82

4.4 Quality of results

This section examines the precision, recall, *f* measure, and accuracy of the classification models for the movie and airline reviews datasets (see Tables 1 and 2). In both datasets, our proposed system's accuracy and *f*-measure outperform other algorithms. In both datasets, our proposed system's accuracy and *f*-measure outperform other methods. As a result, RF and our system have higher accuracy and recall values for the movie review dataset. Our system, RF, and DT are the highest ranked models for movie reviews and airline datasets in terms of accuracy, with more than 91%.

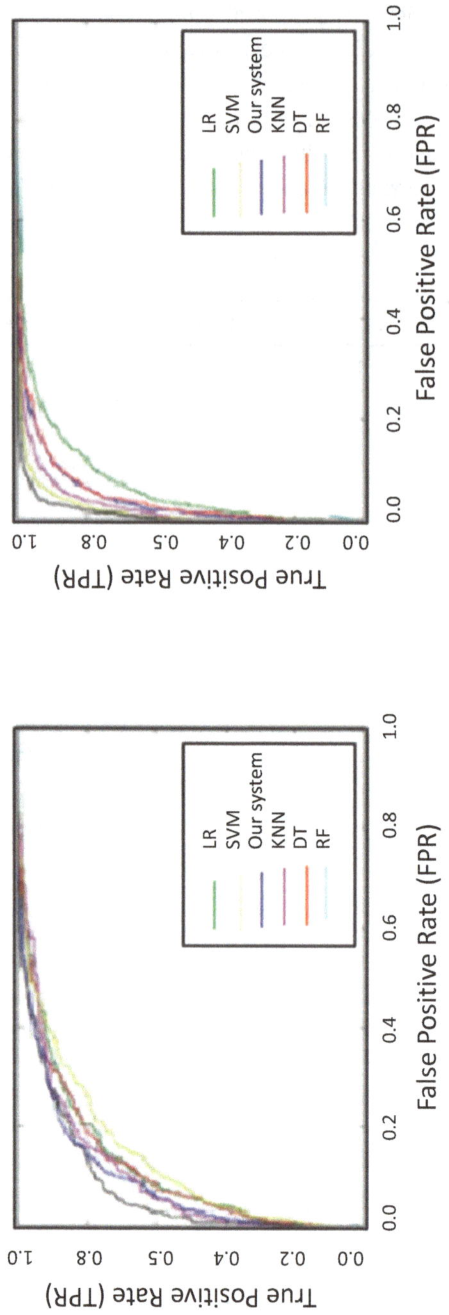

Figure 4: ROC performance comparison for different classifiers with (a) for the IMDB dataset and (b) for airline dataset.

5 Conclusion and future work

This study focuses on the possibilities of employing SA in web 2.0 and 3.0 contexts to obtain consumer sentiments that may be utilized as a proxy for demand. This process allows a company that is far away from a market to fully understand that market and make better supply decisions. We show that it is possible to better understand a large group of consumers. As a result of the content of the consumers' comments, a firm will be able to make better supply management decisions. On the other hand, we examine how SA can be used to evaluate consumer-based massive social media posts in a way that gives supply chain managers a strategic advantage. We present a system that measures customer satisfaction and finding consumer opinions, feelings, and behaviors toward certain topics. When evaluating the effectiveness of the proposed methodology, we employ classifier comparison accuracy, AUC, and ROC. Finally, the results are classified into three classes which are positive, negative, or neutral. It is our intention to use our methodology in the future to construct an analysis system that allows us to gather customer feedback about multiple elements or entities (e.g., items, software products, services, and corporate divisions) on different social platforms using our methodology. Moreover, to further improve the quality of the ML result, we will use different algorithms, which have better accuracy and less computational cost, other than used in this research work.

References

[1] Dai Z, Aqlan F, Gao K, Zhou Y. A two-phase method for multi-echelon location-routing problems in supply chains. *Expert Systems with Applications*, 115:618–634, 2019.

[2] Awasthi P, Dubey N, Sangle PS. Contemporary challenges in crm technology adoption: A multichannel view. *International Journal of Electronic Customer Relationship Management*, 8(1–3):51–71, 2014.

[3] Harris JK, Mueller NL, Snider D. Social media adoption in local health departments nationwide. *American Journal of Public Health*, 103(9):1700–1707, 2013.

[4] O'leary DE. The use of social media in the supply chain: Survey and extensions. *Intelligent Systems in Accounting, Finance and Management*, 18(2–3):121–144, 2011.

[5] Giri KJ, Lone TA. Big data-overview and challenges. *International Journal of Advanced Research in Computer Science and Software Engineering*, 4(6), 2014.

[6] Liu B. Sentiment analysis and opinion mining. *Synthesis Lectures on Human Language Technologies*, 5(1):1–167, 2012.

[7] Agarwal B, Mittal N, Bansal P, Garg S. Sentiment analysis using common-sense and context information. *Computational Intelligence and Neuroscience*, 2015, 2015.

[8] Gursoy UT, Bulut D, Yigit C. Social media mining and sentiment analysis for brand management. *Global Journal of Emerging Trends in e-Business, Marketing and Consumer Psychology*, 3(1):497–551, 2017.

[9] Li Y, Lin Z, Xu L, Swain A. do the electronic books reinforce the dynamics of book supply chain market?"–a theoretical analysis. *European Journal of Operational Research*, 245(2): 591–601, 2015.
[10] Shub AN, Stonebraker PW. The human impact on supply chains: Evaluating the importance of "soft" areas on integration and performance. *Supply Chain Management: An International Journal*, 2009.
[11] Burgess K, Singh PJ, Koroglu R. Supply chain management: A structured literature review and implications for future research. *International Journal of Operations & Production Management*, 2006.
[12] Burgess K, Singh PJ. Using the social system of a supply chain to improve a focal organization's operating performance. *Operations Management Research*, 5(1–2):57–68, 2012.
[13] Aral S, Dellarocas C, Godes D. Introduction to the special issue – social media and business transformation: A framework for research. *Information Systems Research*, 24(1):3–13, 2013.
[14] Rusch E. Using social media in the supply chain. *Retreived December*, 31:2017, 2014.
[15] Biggio B, Roli F. Wild patterns: Ten years after the rise of adversarial machine learning. *Pattern Recognition*, 84:317–331, 2018.
[16] Baryannis G, Dani S, Antoniou G. Predicting supply chain risks using machine learning: The trade-off between performance and interpretability. *Future Generation Computer Systems*, 101:993–1004, 2019.
[17] Zhu Y, Xie C, Wang G-J, Yan X-G. Predicting china's sme credit risk in supply chain finance based on machine learning methods. *Entropy*, 18(5):195, 2016.
[18] Bai C, Rezaei J, Sarkis J. Multicriteria green supplier segmentation. *IEEE Transactions on Engineering Management*, 64(4):515–528, 2017.
[19] Pang B, Lee L. Seeing stars: Exploiting class relationships for sentiment categorization with respect to rating scales. *arXiv preprint cs/0506075*, 2005.
[20] Balahur A, Jacquet G. Sentiment analysis meets social media–challenges and solutions of the field in view of the current information sharing context, 2015.
[21] Moghaddam S. Beyond sentiment analysis: Mining defects and improvements from customer feedback. In *European conference on information retrieval*, 400–410. Springer, 2015.
[22] Tang D, Qin B, Liu T. Deep learning for sentiment analysis: Successful approaches and future challenges. *Wiley Interdisciplinary Reviews: Data Mining and Knowledge Discovery*, 5(6): 292–303, 2015.
[23] Poria S, Gelbukh A, Cambria E, Hussain A, Huang G-B. Emosenticspace: A novel framework for affective common-sense reasoning. *Knowledge-Based Systems*, 69:108–123, 2014.
[24] Folkinshteyn D, Lennon M. Braving bitcoin: A technology acceptance model (tam) analysis. *Journal of Information Technology Case and Application Research*, 18(4):220–249, 2016.
[25] Mikolov T, Sutskever I, Chen K, Corrado GS, Dean J. Distributed representations of words and phrases and their compositionality. In: *Advances in Neural Information Processing Systems*, 3111–3119, 2013.
[26] Bifet A, Frank E. Sentiment knowledge discovery in twitter streaming data. In *International conference on discovery science*, 1–15. Springer, 2010.
[27] Wan Y, Gao Q. An ensemble sentiment classification system of twitter data for airline services analysis. In *2015 IEEE international conference on data mining workshop (ICDMW)*, 1318–1325. IEEE, 2015.

Index

https://doi.org/10.1515/9783110760477-010